Praise for *Dazzle Gradually*

"This is a ripsnorting intellectual barnstorm of a book, a sort of chimeric hybrid of mental genes from Dorion Sagan, his genius mother Lynn Margulis, and his dead father Carl Sagan—surely one of the smartest families on the planet. The book pulls into its mitochondrial group grope such luminaries as Roald Hoffmann, James Lovelock, Ricardo Guerrero, and Eric Schneider—all bona fide members of America's intellectual aristocracy. The result is a remarkably coherent and blazingly original proposal for the next grand narrative of our civilization (now that we have pretty much burned out the Cartesian one)."
— Frederick Turner, author of *Natural Classicism* and *The Culture of Hope*

"Brilliant and fascinating, *Dazzle Gradually* unrolls for us the scroll of life on Earth. These essays show us the intricate complexities of microbes, an atmosphere that performs self-maintenance, our own minds. Margulis and Sagan do not blink at the big questions or hard answers, and their writing is lively, precise, entertaining, and provocative; their passion for science everywhere evident and persuasive. Anyone who has ever wondered where we came from, who we are, and where we may be headed will delight in this extraordinarily exciting book."
— Kelly Cherry, author of *Hazard and Prospect: New and Selected Poems*

"*Dazzle Gradually* is like an air-raid siren, calling for science to reinvent itself for the 21st century; to look beyond the categorization and characterization of things and the traditional view of nature into a highly networked and involved view. In particular, it advises us to descend from our throne of delusion and realize that humanity (with all its technological and cultural trappings), is intimately and inextricably immersed in this grand system along with the protoctists and bacteria, plants and animals, the living world that surrounds us. Our very identity—our minds and souls—are a result of the evolving experiment we call nature. *Dazzle Gradually* is like opening the door to a vast and brilliant garden, which slowly assimilates us as we become part of nature's teaming, humming, growing, and unendingly magical realm."
— Stephen Miles Uzzo, Ph.D., New York Hall of Science

"Biological phenomena are usually viewed in terms of plants and animals. Margulis and Sagan look at them from their extremes: Gaia—the living system of the Earth as a whole—and bacteria. Both Gaia and bacteria dazzle the reader accustomed to conventional fare. It is re-viewing of this kind that paves the way for real advance in science."
— John B. Cobb, Jr., Professor Emeritus, Claremont School of Theology

"*Dazzle Gradually* sparkles with insight and wit as it delves into a host of topics in biology and ecology, linking them in new ways that highlight scientific understanding and speculation at their enjoyable best."
—Donald Goldsmith, co-author of *The Search for Life in the Universe*

"In *Dazzle Gradually*, Margulis and Sagan effectively tap into the cultural waveform through a series of original science essays and provocative ideas to reveal why we are living in an open social networked world, and why survival of the fittest no longer means fit to kill, but fitting in with the rest of life. Simply said, Darwin is left in the dust."

—Mary McGuinness, Co-Director, Sputnik Observatory

"*Dazzle Gradually* invites the reader to push aside the traditional categorization of Earth's biota as an interdependent collective of discrete, membrane-bound organisms and look at it instead as a semi-continuous spectrum of interactive bacterial nations. Like a war correspondent's gritty reports from the dangerous frontiers of evolutionary biology, Margulis' latest reassessments once again threaten to alter our perspective, not only on the life that immediately surrounds and invests us, but on the origin and significance of biological existence within our hydrogen-mediated, thermodynamic cosmos. Cushioned by Sagan's classical allusions and philosophical underpinning this is a book that will seduce, nourish and inspire hungry readers who already know a little of these things."

—Reg Morrison, author of *The Spirit in the Gene*

Dazzle
Gradually

Dazzle
Gradually

REFLECTIONS ON
THE NATURE OF NATURE

LYNN MARGULIS AND DORION SAGAN

 A Sciencewriters Book

CHELSEA GREEN PUBLISHING
WHITE RIVER JUNCTION, VERMONT

 A Sciencewriters Book

scientific knowledge through enchantment
Sciencewriters Books is an imprint of Chelsea Green Publishing. Founded and codirected by Lynn Margulis and Dorion Sagan, Sciencewriters is an educational partnership devoted to advancing science through enchantment in the form of the finest possible books, videos, and other media.

Project Manager: Emily Foote
Copy Editor: Nancy Ringer
Proofreader: Susan Barnett
Sciencewriters Editorial Associate: Dianne Bilyak
Indexer: Marc Schaefer
Designer: Peter Holm, Sterling Hill Productions

Printed in the United States of America
First printing, August 2007
6 5 4 3 2 1 07 08 09 10

Library of Congress Cataloging-in-Publication Data
Margulis, Lynn, 1938-
  Dazzle gradually : reflections on the nature of nature / Lynn Margulis and Dorion Sagan.
    p. cm.
  Includes bibliographical references and index.
  ISBN 978-1-933392-31-8 (alk. paper)
  1. Science—Philosophy. 2. Evolution (Biology) 3. Microorganisms—Evolution. 4. Sex (Biology) 5. Gaia hypothesis. 6. Symbiosis. 7. Sagan, Carl, 1934–1996—Influence. I. Sagan, Dorion, 1959– II. Title.

  Q175.M3626 2007
  501—dc22

                                                              2007021966

Chelsea Green Publishing Company Post Office Box 428
White River Junction,VT 05001 (802) 295-6300
www.chelseagreen.com

A listing of the original places of publication for the essays is on page 258.

Tell all the truth but tell it slant—
Success in Circuit lies
Too bright for our infirm Delight
The Truth's superb surprise
As Lightning to the Children eased by explanation kind
The Truth must dazzle gradually
Or every man be blind

EMILY DICKINSON
1830–1886
Amherst, Massachusetts

To the impersonal principle of
freedom of thought

contents

Foreword by Roald Hoffmann | xi
Acknowledgments | xv

PART I — MNEMOSYNE | 1
1. Red Shoe Conundrum *Lynn Margulis* | 3
2. Truth of My Father *Dorion Sagan* | 8
3. The Uncut Self *Dorion Sagan and Lynn Margulis* | 16

PART II — CHIMERA | 27
4. Power to the Protoctists *Lynn Margulis* | 29
5. Prejudice and Bacterial Consciousness *Lynn Margulis* | 36
6. All for One *Lynn Margulis and Dorion Sagan* | 42
7. Speculation on Speculation *Lynn Margulis* | 48
8. Spirochetes Awake: Syphilis and Nietzsche's Mad Genius *Lynn Margulis* | 57
9. From Kefir to Death *Lynn Margulis* | 70
10. Welcome to the Machine *Dorion Sagan and Lynn Margulis* | 76
11. The Transhumans Are Coming *Dorion Sagan and Lynn Margulis* | 89
12. Alien Enlightenment: Michael Persinger and the Neuropsychology of God *Dorion Sagan* | 101

PART III — EROS | 109
13. The Riddle of Sex *Dorion Sagan and Lynn Margulis* | 112
14. An Evolutionary Striptease *Dorion Sagan* | 121
15. Vive la Différence *Dorion Sagan* | 140
16. Candidiasis and the Origin of Clowns *Dorion Sagan with Lynn Margulis* | 146

Part iv — Gaea | 153

17. The Atmosphere, Gaia's Circulatory System *Lynn Margulis and James E. Lovelock* | 157
18. Gaia and Philosophy *Dorion Sagan and Lynn Margulis* | 172
19. The Global Sulfur Cycle and *Emiliania huxleyi Dorion Sagan* | 185
20. Descartes, Dualism, and Beyond *Dorion Sagan, Lynn Margulis, and Ricardo Guerrero* | 195
21. What Narcissus Saw: The Oceanic "Eye" *Dorion Sagan* | 208
22. The Pleasures of Change *Dorion Sagan and Eric D. Schneider* | 223

Readings | 238
Figure and table list | 245
Index | 246

foreword

We walked into a Monet-land of blues and yellows . . .

Abe Penzer, a wonderful biology teacher at Stuyvesant High School in New York City, knew how to pull us into the world of microorganisms. He began innocently enough, asking us to look under a microscope into what appeared to be a plain drop of water. And there, swimming, its heart beating in its transparent beauty, was *Daphnia pulex*, a water flea. The fact that we could also just see *Daphnia* with our naked eye was important; this seeing was the bridge between the "real," macroscopic world, and what the microscope showed in such splendid detail.

Next, Mr. Penzer gave us some sludge. He couldn't afford pond-water, he said. Sludge more than sufficed—there were *Paramecium*, *Euchlanis*, *Vorticella*, and countless other beings. I remember waiting for a *Vorticella* swarmer to break off. In vain.

As Lynn Margulis' favorite poet writes:

> Faith is a fine invention
> When Gentlemen can see –
> but Microscopes are prudent
> In an Emergency.

Years passed. Perceived pressure to become a doctor turned me away from biology. It took 40 years, and a strange setting, to show me again the wonder of microbiology: the caldera of the Uzon volcano in Kamchatka. We walked off rattling Vietnam-era helicopters, and into a Monet-land of blues and yellows. Up close, there was pool upon pool, one crystal clear, one on the way to orange, bubbles plopping threateningly through the mud clay of another.

And life, in every shade but green! For this wasn't the photosynthetic world—my silver rings turned black from hydrogen sulfide. Water bubbled up boiling at 95°C; pH paper made out that water acid as nitric acid, elsewhere drain-cleaner basic; I'd not put my finger in it were it cool. Round each pool, life—dull red, yellow, beige mats of bacteria, archaebacteria.

Some like it hot. Some want O_2, some do not. This niche just came to be, with a bang. The rest—evolution's game, bricolage. Give it time, hazard, and from the atoms C, N, H, O, S, and metals, life finds its way; the denizens of those colorful mats—in a hell of acid and heat (to us)— find a dear place to play, and pry survival out of a few genes. *Pyrolobus fumarii* grows best at 113°C.

No one would have flown me to Uzon when I was 18. But had I read Lynn Margulis and Dorion Sagan then, I would have become a microbiologist.

For in their books they open up for us a world of microbial wonders. With the time the world gave microbes (in the past), they evolved to use every niche, every source of atoms, energy and electrons, on Earth. Curiously, the infinite variety of life that evolves not only gives splits and sunders, those thousands of beetles. It also interweaves biota with Earth, waters, and atmosphere to produce—in the grandest of symbioses—an emergent form, a superorganism, Gaia.

Dr. Margulis is what science, ossifying as it becomes institutionalized, needs. She has absolutely no "Faith," in the sense of those "Gentlemen" in Emily Dickinson's poem. So, she is the ultimate doubter (though, boy is she sure her ideas are right; I'm smiling). More than that, Lynn's brilliant mind has found the road between curiosity and speculation that so many scientists have allowed to grow over.

Lynn is the mistress of the creative hypothesis, the great speculatrix. In this wonderful book you will find her hypotheses in abundance. You will read of mind processes originating from "an unholy microscopic alliance between hungry swarming killer bacteria and their potential archaebacterial victims." She suggests that the mitotic cells of all nucleated organisms harbor spirochete bacterial remnants. She postulates a symbiotic mechanism for the evolution of our sensory cells. Dorion speculates about the etiology of clowns.

I love Lynn's stories. And many of them have turned out to be right.

Like mother, like son? Not quite. It's not easy to take them apart, for they often write together (an action in my experience usually guaranteed to separate people). Dorion is the passarelle, the gentle bridge between his mother's science and the magic of science fiction (science fiction writers should read this book, it has material for a library). He is a gifted expositor, ever curious, drawing us into his exciting encounters with science and ideas.

I have fun trying to guess which ideas are Lynn's and which are Dorion's. But it doesn't matter; what they have created in *Dazzle Gradually* is a land twixt biology and imagination. Not everyone will follow them—I don't like their explicit support of Peter Duesberg's hypothesis of a non-HIV cause of AIDS, nor do Michael Persinger's investigations of the effect of electromagnetic fields on the psyche pass the "smell right" test to me. But I have absolutely no trouble accepting these variant views as part of the Margulis/Sagan gestalt. Or, since I'm using a German word, perhaps it should be "Gesamtkunstwerk"—a synthesis of the arts.

In *Dazzle Gradually* we have one of the great iconoclastic biologists of our time, and her son, both excellent writers, firing ideas at us, reflecting, asking questions, making connections. "Truth's superb surprise" is their gift to us.

ROALD HOFFMANN
Ithaca, New York, 2007

acknowledgments

These essays are grounded in the deciduous forests, university campuses, and cities of the northeastern United States, mainly Boston, Woods Hole, and Amherst, Massachusetts. In those places and others, we have conversed with myriad students, scientists, scholars, naturalists, philosophers, social critics, editors, and businesspeople. Among those whose work and ideas particularly influenced ours are Elso S. Barghoorn, David Bermudes, Daniel Botkin, Michael Chapman, the members of the Commonwealth Book Fund Committee, Michael Dolan, Ricardo Guerrero, John Hall, H. D. Holland, G. E. Hutchinson, Wolfgang E. Krumbein, Antonio Lazcano, James Lovelock, Heinz Lowenstam, David Luck, Alan McHenry, Mark McMenamin, Harold Morowitz, Kenneth Nealson, Dennis Searcy, Eric D. Schneider, Lewis Thomas, James W. Walker, and Peter Westbroek.

For manuscript preparation, societal, and creative support we thank Celeste Asikainen, Connie Barlow, Andrew Blais, Gérard Blanc, Howard Bloom, Ron Blum, Mads Brugger, Lois Brynes, Peter Bunyard, Chris Carlisle, Emily Case, Carmen Chica, Kathryn Delisle, Joanne DeLuca, James di Properzio, Michael Dolan, Sona Dolan, Sean Faulkner, Rene Fester, Gail Fleischaker, Deborah Fort, Carolina Galan, Teddy (Edward) Goldsmith, Steve Goodwin, Aaron Hazelton, Jeremy Jorgensen, Rita Kolchinsky, Tom Lang, Janine Lopiano, Christie Lyons, James MacAllister, Jennifer Margulis, Zachary Margulis-Ohnuma, Humberto Maturana, Mary McGuinness, Claude Monty, Tonio Sagan, Bruce Scofield, Lorraine Olendzenski, Michael Persinger, Mercé Piqueras, Russell Powell, Donna Reppard, Brian Rosborough, Michael Stone, William I. Thompson, Sonya Vickers, Constanza Villalba, Jorge Wagensberg, Peter Warshall, and Janet Williams.

We are grateful to Bruce Wilcox of the University of Massachusetts Press, William Frucht of Basic Books, and Lewis Lapham, former editor-in-chief of *Harper's Magazine*, for encouragement. We received some financial support from the Alexander von Humboldt Stiftung, the Tauber Fund, and the University of Massachusetts College of Natural Sciences and Mathematics. We thank our former agents John Brockman and Katinka Matson, our present agent, Georges Borchardt, and especially Margo Baldwin, John Barstow, Emily Foote, Abrah Griggs, Collette Leonard,

Jonathan Teller-Elsberg, Nancy Ringer, Shay Totten, and others of the wonderful team at Chelsea Green who bestowed upon us the unexpected literary gift of our Sciencewriters Books imprint. Judith Herrick Beard (Typro) aided in innumerable intelligent ways to facilitate manuscript preparation. Dianne Bilyak served for months with alacrity and diligence as editorial assistant to Sciencewriters.

The Sciencewriters Partnership, from which Sciencewriters Books sprang, began in the spring of 1981, when we were visited in Lynn's Boston apartment by a colorful character in a pimp hat and three-piece suit, the literary agent John Brockman. As the host of an intellectual salon called the Reality Club, John sought out the most "cutting-edge ideas in science," and he had come to persuade Lynn to popularize her work on symbiosis. She adamantly refused; she was strictly an academic and did not even read newspapers or watch television, let alone write popular books. "Go ask my ex-husband," she said. "That's his specialty."

At this point Dorion, then a college senior, launched into a monologue, claiming among other things that extraterrestrial intelligence was merely a replacement for religion in a secular age.

"That's good," said Brockman. "Why don't you write a book? You can write about growing up with your father."

"Okay," Dorion said. "But I'll have to make it all up because my parents separated when I was three."

"Fine," said John. "As long as it's nonfiction."

Thus sprouted Sciencewriters, which has grown under the green thumb of Margo Baldwin, fearless, peerless head of Chelsea Green Publishing Company, into the present corporately unbeholden imprint.

The scientific work in the laboratory of Professor Margulis was supported from 1972 through the end of the century by NASA (Drs. Richard Young, John Rummel, and Michael Meyer often fought to make this possible). Financial support for students, scientific colleagues, and curricular materials that came from the Richard Lounsbery Foundation, New York was due to help from Lewis Thomas, Alan McHenry, and especially Marta Norman. This support, which lasted from 1983 until 1994, enabled our books, films, and science-education materials. Aid from the Boston University Graduate School, and the University of Massachusetts at Amherst is gratefully acknowledged.

Finally, we thank Robert Hutchinson for putting us in touch with Bill Atkinson, whose spectacular digital photograph of a tiger's eye stone from the 3.63 billion-year-old Marra Mamba iron formation in Western Australia happily adorns the cover of this book.

Dazzle
Gradually

part one

mnemosyne

The mythological mascot who oversees this section is Mnemosyne, the goddess of memory and mother, some say, of the Muses, nine nymph goddesses of art, science, and inspiration, including Calliope, the muse of eloquence, Clio, the muse of history, and Euterpe, the muse of music. Impregnated by Zeus, king of the gods, Euterpe gave birth at the base of Mount Olympus. We chose this wonder woman because the first two essays in this short section are time-based, personal, and reflective. We might have chosen Cronos, the Titan father of Zeus, because he too is associated with time. But Cronos, sired of incest between Gaea and Uranus, devoured his children. Cronos

would have swallowed Zeus, too, had not Mother Earth—
Gaea—counseled his mother, Gaea's daughter-in-law Rhea
(whose name means "flow"), to switch her baby for a rock
covered in swaddling clothes. Zeus thus survived to come
of age and ultimately had his way with Memory.

Here in these first essays we introduce ourselves, in a
scary and personal way, to the reader. Lynn and Carl for-
mally married in a ceremony that pleased her mother
(Leone Alexander, wife of Morris Alexander, a prominent
activist Chicago lawyer and owner of Permaline, a company that installed plastic stripes on roads and highways;
he later became an assistant state's attorney for Illinois).
Lynn was nineteen years old and Carl twenty-four
(chapter 1, figure 1.1). Lynn had been sharing her life
with, on and off, and learning science from Carl since she
had met him at the age of sixteen.

In the first chapter Lynn explains her real feelings about
women and the scientific career. In the second chapter
Dorion tells you of his peculiar experience as the eldest of
Carl Sagan's five children, one per decade. The third essay,
on the self, begins the transition to the concept of
chimera—the real, biological self that combines multiple
beings. The self is never the Platonic ideal implied by the
word.

Red Shoe Conundrum

LYNN MARGULIS

My honest opinion about children, mothers, mates, marriage, and the pursuit of science. These passions are more personally pursued in my novel *Luminous Fish: Tales of Science and Love.*

For as long as I can remember, when someone asked me what I wanted to be when I grew up, I answered, "An explorer and a writer." Explorer of what? As a child, I didn't know: undersea cities, African jungle pyramids, unmapped tropical islands, polar caves. "Whatever will need exploring," I said without hesitation. Today, nearly incessantly, I explore with passion the inner workings of living cells to reveal their evolutionary history. And as soon as I learn something new about bacteria or insect symbionts that helps explain the history of life on the Earth's surface, I write about it.

So you see, I am, after all these years, an explorer and a writer. Science for me is exploration, and no scientific work is complete if it has not been described and recorded in an article by the scientist herself (the "primary literature") or in a book or paper by someone else (the "secondary literature"). Much of my day is spent in description: generating papers that speak to fellow scientists and graduate students, talking in classes or lecturing to amuse the curious, writing notes and observations, collecting references, and jotting down the insights of others. I have become a mother (of four), quit my job as a wife (twice), become a grandmother (seven times, so far), and for the past twenty-four years been a transatlantic partner of a father of three.

Because no one in my early life ever even mentioned the existence of science, I never realized until adulthood that I could participate in its great adventure as a profession. Unlike many friends, neither as an adolescent nor as a young adult did I wait for "my prince to come." Rather I expected some—any—opportunity to join serious expeditions. Then, as today, I read nearly everything in sight: bottle labels, train schedules, recipes, Spanish poetry, and novels. Decades ago, on the south side of Chicago, I used to ride the "IC" (Illinois Central Railroad) some forty

Figure 1.1 June 16, 1957, Chicago. Lynn's mother, Leone Alexander; Carl Sagan; Lynn Margulis.

minutes, in both the stifling heat of summer and the freezing cold of winter, at least once weekly to the downtown Loop for ballet. Ballet classes (demanding, exhausting, French, and irrelevant) were sufficiently escapist to be captivating before scientists or exploratory missions entered my life.

CHOICES

One film moved all of us in those days: we all idolized redheaded Moira Shearer prancing in *The Red Shoes*. Set near Nice on the Mediterranean, close to a place with a marine station (Villefranche-sur-Mer) that I got to know many years later, this romantic movie mesmerized my dancing classmates. The talent of this beautiful ballerina in the prima donna role was exhilarating, as was her true love for her sexy, handsome beau. I remember feeling anger at the melodrama of that movie, however. I thought the dichotomy of her life that led to her self-instigated fate utterly ridiculous.

Why did there have to be "necessity to choose" between devotion to a man and devotion to a career? What generated the psychic dissonance that distracted her to destruction? Obviously there was no reciprocity: if the star had been male, he would not have been driven to choose. He simply would have taken a wife. Instead, under relentless pressure to be the perfect dancer whose shoes run away with her, the ballerina yields to the dance master's demands that she remain in the spotlight, stage center of his world. But, equally enamored of her man, she is driven by another exigency: her lover demands that she marry him and have a family.

Why hadn't she simply married her lover, borne her children, and continued dancing? Hollywood resolved her dilemma tragically, making the young heroine jump to her death from the summit of a seawall. What infuriated me was the idea that the healthy, beautiful, and ambitious ballerina had to accept the either-or notion imposed upon her by the two men who ran her life. Had she simply opted for everything, however, she would have deprived the film of its trumped-up fatal conflict. Wasn't a strong family life and a career possible for Moira Shearer's character? Isn't such a full life even easier today in the age of food storage by deep freeze, the private automobile, the dishwasher, and the laundry machine?

At age fifteen I was certain that the ballerina died because of a silly, antiquated convention that insisted that it is impossible for any woman to maintain both family and career. I am equally sure now that the people of her generation who insisted on either marriage or career were correct, just as those of our generation who perpetuate the myth of the superwoman who simultaneously can do it all—husband, children, and professional career—are wrong.

Today many students, especially women, ask me for enlightenment on how to combine successfully career and family. When they learn I have four excellent, healthy, grown children and never abandoned science for even a single day in over forty-five years, they request my secret. Touting me as an example of an American superwoman, they label me a "role model" (a term I despise). But there is no secret. Neither I nor anyone else can be a superwoman.

Aspiring to the superwoman role leads to thwarted expectations, the help-less-hopeless syndrome, failed dreams, and frustrated ambitions. A lie about what one woman can accomplish leads to her, and her mate's, bitter disappointment and to lack of self-esteem. Such delusions and self-deceptions, blown up and hardened, have reached national proportions. Rampant misrepresentation of feasibility abounds as everyone falls short of the national myth peopled with a happy family, educated children, and professionally fulfilled parents. Something has to give: the quality of the professional life, of the marriage, or of the child rearing—or perhaps all—must suffer.

The unreality of such expectations, coupled with the gross inadequacy of our educational system, such as it is, often leads to despair temporarily relieved by mind-numbing video games, drugs—marijuana, whiskey, cocaine—or other escapes.

Each husband, wife, and child in this sea of false hope suffers the crushing pain of inadequacy. In the United States we value the beauty and strength of youth, but as a culture we disdain love for children as "touchy-feely" and denigrate homemaking as trivial and unworthy. We marginalize or expel the elderly and ridicule life on communes. By no means are the homeless on the street the only ones without homes. Unwilling to care for our greatest resource and those in direst need—our infants and children—we, speaking through money, debase their instructors, despising the seriousness needed to acquire a fine education. Our culture laughs at the inquisitive while it lauds the merely acquisitive.

I have not in any way overcome these stresses or resolved these problems. I have just ignored them, as if they were laws that do not apply to me. Looking beyond such social heartaches, I chose intellectual exploration as my way of life and allied myself with nonhuman planetmates. I chose the scientific quest, rather than devoting myself to an arbitrary integrity of family and human community.

And, of course, I never jumped off the ballerina's cliff; the thought of abandoning life itself has always been unthinkable. Be warned, though: I do not offer a recipe for personal fulfillment. Superwoman does not exist, even in principle.

Mine is the story of scientific enthusiasm and enlightenment coming to a foolish and energetic girl who turned down dates on Saturday nights and who never watched television. The point is that I was willing to work. This is not a statement of advocacy, as no single answer or easy path suits every woman. Probably I have contributed to science because I abandoned two husbands and many more boyfriends. I failed to answer hundreds of telephone calls and came late to many parties. Although remiss in many social activities I chose to stay with the children. I've been very poor, but I've never been sorry.

Children, husband, and excellence in original science are not simultaneously possible. Yet I fervently believe that those few women who feel the urge must be encouraged to pursue and maintain their scientific careers. Such women need our help. If life does not pose its problems as melodramatically as a Hollywood movie, neither does it resolve them so cleanly or definitively.

Yes, women can, of course, be superb scientists, but only at great sacrifice to their social lives and its obligations. Most critically productive women and girls must be surrounded by supportive and loving men and boys. We all need a cultural infrastructure that respects the deep needs of our young children and older family members. Let us hope that the provision of such enablers as scholarship monies, family-leave opportunities, enlightened health insurance programs, imaginative and indulgent day care for preschoolers, and afterschool play programs will increase the probability that talented and determined young women will contribute much more to the future scientific adventure than they ever have managed to in the past.

Truth of My Father

DORION SAGAN

This essay, originally published as "Partial Closure" in an issue of *Whole Earth* that also had a memorial of Timothy Leary by his son, was pretty much all that could occur between Carl and Dorion after Carl's death. Carl became a better father with each of his successive children, but he failed my sons. (LM)

MEMORY OF MY DAD

Of the several cartoons in which my father's image has appeared, perhaps the most famous, published in the *New Yorker*, depicts two aliens coming to Earth. "No, not Carl Sagan," says one of the saucer-bound spacelings, "too hokey. Let's grab somebody less obvious." He will forever be associated in the popular imagination with the cosmic, the extraterrestrial, the postreligious scientific sublime. At six feet two inches, with a bass voice (I heard it in the womb), perfect diction, an encyclopedic memory, *um*-less speech, and a preternatural (if to me, at times, privately aggravating) way of orating reasoned paragraphs that made other people's speech sound like illogical jabberwocky, he was—and is—larger than life. It is said that people's weaknesses are their strengths. I guess my problem with him boiled down to this: just as talking familially to the masses about the beauty of the cosmos on the *Tonight Show* with Johnny Carson made him a father figure to millions, so his air of unassailable authority, 1950s-style paternalism, and intellectual arrogance made him emotionally distant as a father. He was that same television guest in his own living room, only he was now also the host. If I, his eldest son, was a privileged member of his audience, I was still, well, a member of his audience. Sometimes one felt more like a camera than a family member. On one hand, he was always "on stage," perhaps even to himself. On the other hand, we did have some thrilling conversations and it was a rare privilege to be in his world-class company. When I was twelve, listening with great frustration to conservative talk-show host Avi Nelson in Boston, I used to daydream that my father would call in and put the smarmy rhetor in his place—blow him away with reason.

My father is said to have been the most recognized scientist in the world during his time.[1] * As a teenager I had friends who told me to tell him to run for president. People I'd never met proclaimed to me that he was the smartest person in the world. A woman who later became betrothed to a millionaire tycoon of a girlie magazine empire first sent naked pictures of herself to my father in hopes of sparking a non-other-worldly interest. No doubt much of this was due to his naturalness on camera, his telegenic presence that was showcased in the very successful PBS television series *Cosmos*. He was a passionate defender of the truth as he saw it, revealed by the scientific method. And he was a good scientist. He postulated that Venus was so hot because the carbon dioxide in its atmosphere had led to a runaway greenhouse effect; this was later confirmed. Although he would have loved finding life on Mars, he theorized that the changing surface of the red planet was due not to seasonal vegetation but to violent dust storms. His theory not only was proved true but also provided the starting point for the notion that a similar dust-raising, sun-obscuring nuclear winter could threaten Earth's agriculture and life on a global scale. Any historical account of the end of the Cold War might ascribe a role, perhaps a pivotal role, to the dissemination of this theory. And he showed that brownish substances similar to those found on Jupiter and its moons could be synthesized in the laboratory; unfortunately, these organic compounds, called tholins, probably contributed to his early death, in 1996 at the age of sixty-two, by leukemia.

Considering that my mother and he split when I was so young, I was secretly gratified when his career later took a turn from the extraterrestrial to the worldly. He and his third wife, Ann Druyan, were arrested at a Nevada nuclear test site, protesting nuclear arms policy. He took his message and his indefatigable reason to the floor of the Senate; he successfully debated Secretary of Defense Caspar Weinberger on national TV; he refused three invitations to the Reagan White House; he showed that the power of the intellect and moral authority are still forces to be reckoned with in this political age of sound bites and sloganeering. At a memorial service at the Cathedral Church of Saint John the Divine in Manhattan shortly after Carl's death, physicist Roald Sagdeev, a former adviser to Mikhail Gorbachev and director of the Space Research Institute in

---

*Numbered notes are at the end of each chapter. Full references are listed in *Readings* beginning on page 238.

Moscow, credited my father with ending the Cold War. If, I reasoned at some unconscious level, I was to be passed over for the sake of his career, it was more pleasant to imagine I had been set aside for world peace than for exobiology (one of those rare disciplines that, like parapsychology, is a science still without an object of study).

THE WHITE KNIGHT OF SCIENCE

My brother Jeremy, who shared the podium with Vice President Al Gore at the memorial service at Saint John the Divine's, described Carl as a "noble truth teller." I do believe this is largely true and the source of much of my father's authority. He had an exquisite integrity and thirst for knowledge; he was in love with science and the search for truth. He was also, I suspect, driven to seek perhaps unrealistic levels of clarity due to his discomfort with the complexity of interpersonal machinations. Not only was he born in 1934 in the depressed economy leading up to World War II, but his brilliant yet frustrated and overprotective mother, Rachel, who grew up in the Jewish slums of Vienna, was no stranger to trickery. The deviously brilliant wordsmith had for decades deceived her treasured son by calling mushrooms "onions" and onions "mushrooms." The linguistic travesty, whose purpose was to ensure Carl consumed the fungi he liked despite his subtly evident initial distaste for their name, was tragicomically revealed when, upon ordering for himself the culinary delicacy as a young man at a restaurant, the waiter brought onions, for which Carl cared not. I consider this true tale an index of the prison of received wisdom in which the astronomer-to-be found himself and of the escape route that science, with its satisfying mystery-destroying method, could provide. When one's own mother was capable of expending so much energy to keep such a minor secret, I theorized, one's trust might find refuge beyond any individual in the methodology of science, in which all statements have to be vouchsafed, verified, checked against the evidence of reality. I remember him telling me of his thoughts that his mother was a secret drunk—because he had discovered beer in the bathroom. In fact it was a peasant trick. To rinse one's newly washed hair in the yeasty brew was said to thicken and smooth it. Who could forgive him for being suspicious? When I learned magic tricks and as a teenager performed a little close-up show for him, he laughed delightedly. Later, he went to the Magic Castle in Hollywood and became privy to a few methods by which magicians amaze their spectators. What shocked him, he said, were the lengths to which magicians would go to obtain their effects. For him the contrast between humans, and the devious things we do, and nature, and the reliable way it works, could not have been starker: "Nature doesn't cheat," he told me on more than one occasion. And yet wasn't it true that humans emerged from nature? And isn't science human? Carl also learned after Rachel died that he had an uncle that he didn't know even existed. Rachel's younger half-brother was homosexual. That he had been lobotomized and institutionalized as "treatment" Carl learned at Rachel's funeral. She took the embarrassing secret to her grave. Thus I wondered whether Rachel might have cultivated his need for truth—and exacerbated his faith in science's ability to find the final truth of a nature that "doesn't cheat."

Many of our most intense discussions were epistemological. But I find it interesting, although it is not quite fair since he is no longer around to defend himself, to see where my father and the truth were at variance— not only because it shows his humanity, but also because it touches upon some of the weaknesses of the positivist tradition, to which he gave such an eloquent, consistent, and ardent voice.

The avidness with which my father attempted to protect the hallowed realm of science from the encroachments of pseudoscience was admirable. The U.S.A. is, after all, an anti-intellectual country. Our greatest contribution to philosophy is pragmatism—or, in the words of the sneaker goddess Nike, "Just do it." Had my father called into question the attitude of the *Skeptical Enquirer* (on whose board he served), with its we-are-the-knights-of-reason philosophical naïveté, as avidly as he did that of the *National Enquirer*, with its we-don't-print-lies-we-just-believe-anything-anybody-tells-us stories about aliens and astrology, I suspect he would never have attained the position of moral and scientific authority that he did.

He hated it when I claimed with Nietzsche that nature does not admit of any absolute objectivity but is already always an interpretation. (And I must admit, when I said such things, I was partly playing the devil's advocate, testing and tweaking him to persuade me that the poststructuralists were wrong.) He hated it when I spoke of the metaphorical nature of all language, including scientific discourse. Or when I pointed out the rhetorical way in which he used words like *science* and *evidence*. Don't get me wrong: I think science packs as powerful a punch as most philosophy, and I agree that science's habitual appeal to nature gives it the upper hand. And yet, as I tried to tell Dad, science's brilliant practice of keeping its truths provisional and open to revision in light of new evidence tends to make it cocky. Scientists think they are not only above superstitious pseudoscience but also beyond any obligation to examine the all-too-human philosophical roots of scientific practice.

THE DEMON'S HAUNTS

The final book to appear by him while he was still alive, *The Demon-Haunted World*, dedicated to my rapper son Tonio, was a heartfelt critique of intellectual-fraud wishful thinking; a defense of science, reason, and the wondrousness of nature unelaborated by feel-good fantasy. Recognizing my knowledge of sleight-of-hand magic, whose techniques are sometimes used by the unscrupulous to pretend supernatural powers, and our ongoing philosophical tête-à-tête, my father sent me the manuscript for comment. But when he wrote of those "standard postmodern texts, where anything can mean anything," I was taken aback. To which standard postmodernist texts was he referring? Had he, in fact, read any? There was a huge difference, I said, between a philosophical critique of science—an examination of its historical and social context, its inevitable assumptions, and its limitations—and pseudoscience, the uncritical acceptance of unsubstantiated beliefs. Nor is the second-rate nature of many academics sufficient reason to dismiss philosophical skepticism about science—any more than Christ's message of love should be trashed because he has not yet saved us from belligerent Christians.

I thought it hypocritical of Carl to preserve science from the critical skepticism that it aimed so effectively at the rest of the world. The situation is similar to that of post-Enlightenment philosophers who used reason to deconstruct reason. Needless to say, one cannot be critical of everything. Some things must be taken for granted. But that is part of the point. Behind the shiny chrome facade of twentieth-century science is not the Wizard of Reason but good old fallible human nature.

Not long ago I heard myself speaking in an *um*-lessly low and authoritative voice. It was my father talking to his third wife and production partner, Annie Druyan. I was channeling him:

"Annie, this is Carl. I am making this communication to you through my son, Dorion. The search for extraterrestrial intelligence is not what we imagined. Aliens already exist, in what the Christians call Heaven. Everything is already here, what were known as 'ghosts' in the hydrogen-wavelength spectrum. The aliens are not dwelling on other planets, as the Drake equation suggests. The 'aliens' are us, projected orthogonally against the gradient of linear time. Ironically, Dorion was right. I am going to continue to use his body to project this crucial new evidence to the widest possible scientific and lay audiences. Annie, I'd like you to please call a press conference, tomorrow at four p.m., in the *Cosmos* office."

PARTIAL CLOSURE

My mother tells me that Carl was jealous of her attentions to me when I was a baby—a potential Oedipal drama hardly unique in the 1950s. "I trust that name has served you well," my dad once said. "Well, yeah, Dad, but . . ." At the height of the popularity of *Cosmos*, seen by half a billion people, my evolutionary biologist mother and I received a book contract from Simon and Schuster to write *Microcosmos*, about our bacterial ancestors. My father called *Microcosmos* a rip-off title and warned that people might see my name and confuse my book with his and purchase it mistakenly. I was surprised that he might think that a possibility, but as P. T. Barnum infamously reminded us, "Never underestimate the stupidity of the American public." Animal empathy author Jeffrey Moussaieff Masson recounted at a conference we both attended at DePaul University how his book on elephants had become a best seller: Oprah Winfrey had mistakenly picked it up off a table to flag for her book club. My response to my father after he criticized me for working too much with my mother was that perhaps he and I might work on some manuscript. Alas, it was not to be.

The only reason that I would dare to think that I might temporarily fill his shoes long enough to finish an essay such as this is because I am wearing them! How, you may well ask, could I, at only five feet ten and three-quarters inches and some 150-odd pounds, fill the great Carl Sagan's shoes? The answer is simple. Carl bought these shoes (black zipper-lined Beatle boots, similar to the ones that British eccentric Nicholas Guppy informs me caused quite a stir when Carl wore them to a lecture in the 1960s at the staid British Astronomical Society) in Italy, in the morning. As any proper member of the landed aristocracy knows, you don't purchase shoes in the morning; feet swell during the day, so the shoes may be too tight. Fortunately for me, my dad didn't know this erudite shopping fact and bought three pairs on that happy morning. I became the proud recipient of two of them.

In 1985 my father arrived in the intensive-care unit at my bedside in Syracuse after I was the victim of a violent crime in Florida. He was there for me, and it improved our relationship. Weak and paranoiac before prophylactic brain surgery to forestall the possibility of meningitis, I was assured by him that my Hindu-like fears about the Godhead splintering itself into separate selves, each no longer bored because of the spellbinding illusion of death, were not that creepy or unusual. On the contrary, he assured me, whether or not God could kill Himself was an

ancient question of Western theology. My father believed in the God of
Spinoza and Einstein, God not behind nature but as nature, equivalent to
it. So can God commit suicide? The first law of thermodynamics, that of
the conservation of energy, suggests that God perhaps cannot. Every boy
needs his father.

As time went by my relationship with my dad improved to the point
where I felt I could express my lingering disappointment and anger. After
I dared to do this a little, our relationship began to deteriorate again. I
found myself in heated arguments with him about scientism and capi-
talism. I knew that unresolved emotional problems were firing our intellectual arguments.

As other children before me have discovered, giving up on your
demands, just letting your parent be another person, an adult, is freeing.
Most of us think our parents are superhuman when we're young. Given
my celebrity scientist dad, my illusion was almost realistic. After "giving
up" on him, though, I felt relief. Even he mentioned that our relationship
had improved and asked whether I knew why that was. Basically, I gave
up on you, I admitted. Is there anything he could do to improve our relationship, he asked me. I said no, not really, it's a Zen thing. Nothing that
can be done. Accept it.

But the reconciliation, the closure I thought I had achieved by abandoning my demands, by no longer expecting him to do something that he
really couldn't do anyway—make up for the long-gone past—was ques-
tionable. At the funeral I had prepared something about his life and science, but after reading it to Jeremy, who told me it could have been written
by a colleague, I scrapped it. Instead I gave, at graveside, a veritable public-
service announcement on the benefits of reconciling with your parents
before they die, which implicitly I had assumed I had done. I had not.

A few days later I found myself regretting that I had let him off the hook
so easily. However Zen, telling him there was nothing he could do meant
he did not have to try. In retrospect, a week after his death, I doubted the
wisdom of such nonattachment. I loved him. I wanted to love him.

I wanted him to love me. I cried—for him and all that he was, and was
not, to me. It was clear then that my "sage" speech on the importance of
premortem reconciliation with parents was premature. Practice what you
preach. I wasn't reconciled myself. Yet. But then I had this cool dream: I
am sitting outside, a person or two on the balcony across the street. It
slowly grows into a party with people sitting on the wall of the balcony
over a convenience store and whooping it up each time a car passes by. My

father is on a bench on my side of the street but I don't recognize him until I hear his voice. We have somehow reconciled, I somehow know he is going to die, but there is something strangely weak and soft about him. Later, I go to the convenience store across the street. Somebody buys a confection, gives it to me because his "tongue couldn't fit all the twists." I taste it, it is dry—I discard it. We make an exodus past the store to a field where some of us gather. I wonder where we are going. Someone tells me that "entrepreneurs" are selling stuff that the family has eaten—reliclike, because of my father's fame. Then, beyond a couple of painted clowns scooting by, I see my father leaving. He is going, he is going to die. Now he is almost gone. I hug him and he returns my affection with a truly great loving hug. I find myself on my knees like a toddler, crying and trying to hold on to my dear, departing father. Except he isn't there. I look up and see two great leglike trees, topped by the canopy and the sky. He is gone.

When I wake up, I feel reconciled. My regression into a bawling toddler, and his disappearance back into nature, has somehow done the trick. The trees, I recall, are especially interesting. When I was eight or so my father told us a story that continued over the course of several weekends. Apparently he made the serial up as he went along. It was a change from his usual stories of time travel and black holes, neutron stars and other dimensions. But even in this story there was an astronomical connection: the characters had names like Callisto, Ganymede, Europa, Io, and so on—they were named after the moons of Jupiter. And curiously, in one of the first installments, he said something I will never forget. He mentioned four trees, oak or pine—I can't remember—in a line. Their significance, he said, was to become clear later in the story. He went on to regale us with the twists and turns of this spoken fiction. He never returned to the significance of the trees. I now realize that the trees' mention was a plot device, a suspense builder that made you want him to keep telling the story, if for no other reason than to find out what they meant. And two of those trees, at least, have shown up, their significance rich—they are nature beginning to replace my dad's body—in my reconciliation dream.

Chapter 2 Note

1. For excellent documentation of C. Sagan's life and contribution to science and its communication see Poundstone, 1999.

The Uncut Self*

DORION SAGAN AND LYNN MARGULIS

> Speeches and books were assigned real authors, other than
> mythical or important religious figures, only when the author
> became subject to punishment and to the extent that his dis-
> course was considered transgressive.
>
> MICHEL FOUCAULT, *Language, Counter-Memory, and Practice*

full circle, not based on the rectilinear frame of reference of a painting, mirror, house, or book, and with neither "inside" nor "outside" but according to the single surface of a Möbius strip. This is not the classical Cartesian model of self, with a vital ensouled *res cogitans* surrounded by that predictable world of Newtonian mechanisms of the *res extensa*; it is closer to Maturana and Varela's conception of autopoiesis, a completely self-making, self-referring, tautologically delimited entity at the various levels of cell, organism, and cognition.[1] It would be premature to accuse us therefore of a debilitating biomysticism, of pandering to deconstructive fashion, or, indeed, of fomenting an academic "lunacy" or "criminality" that merits ostracism from scientific society, smoothly sealed by peer review and by the standards of what Fleck calls a "thought collective."[2] Nor would it be timely to label and dismiss us as antirational or solipsist. All such locutions stem from the mundane reason, the ethnocentric conception of self that precisely comes under question here. "The philosophy of the subject," writes Jürgen Habermas, "is by no means an absolutely reifying power that imprisons all discursive thought and leaves open nothing but a flight into the immediacy of mystical ecstasy."[3] On the one hand we position ourselves beyond the sixteenth-century European Enlightenment, its faith in reason, the arrogance of its secular priests, and the later Darwinian smarm. In this sense we have a poststructuralist, postmodern, nonrepresentational view of self.[4] On the other hand we dialectically question this position, motionlessly turning it inside out, as it were,

---

*The full circle of existence; this essay is meant to begin with lowercase "f." If you wonder why just read the end first.

and paying heed to the successes of scientific positivism and biochemical reductionism—movements that philosophically cannot (at least provisionally) be disentangled from the pervasive influence of Indo-European grammar, subject-verb-object structures, and the like. In this sense our view of the organism is less ontological and more biological; the order of metaphysics and physics, the primacy of philosophy over biology, undergoes a reversal more in keeping with the academic notions of self and the anthological effort to enclose in a coherent, comprehensive, rectilinear manner.[5] Membrane-bounded indeed.

But the membrane is no concrete, literal, self-possessed wall; it is a self-maintained and constantly changing semipermeable barrier. The idea of the semipermeable membrane permits us to jump organizational levels, from intraorganismic cell to cellular organism to organismic ecosystem and biosphere. Whether we are discussing the disappearing membranes of endosymbiotic bacteria on their way to becoming organelles or the breakdown within the global human socius of the Berlin Wall, we must revise this rectilinear notion of the self, of the bounded I. Alan Watts pejoratively referred to it as the "skin-encapsulated ego"; indeed, even though so deeply entrenched, this bounded sense of "self" seems to us to be thoroughly natural—it is neither a historical nor a cultural universal. For example, the Melanesians of New Caledonia, known in French as the Canaque, are unaware that the body is an element that they themselves possess; the Melanesians cannot see the body as "one of the elements of the individual." So, too, the Homeric epics never make mention of a body—the flesh-enclosed entity we today take for granted as the definable material self—they speak only of what we would think of as the body's parts, for example, "fleet legs" and "sinewy arms." "The idea of the 'self in a case'. . . ," writes Norbert Elias, "is one of the recurrent *leitmotifs* of a modern philosophy, from the thinking subject of Descartes, Leibniz' windowless monads and the Kantian subject of knowledge (who from his aprioristic shell can never quite break through to the 'thing in itself') to the more recent extension of the same basic idea of the entirely self-sufficient individual."[6]

Psychoanalytically, the sense of self on the level of personhood has been construed to be a convenient fiction, an effect of infantile representation that is jubilant but essentially ersatz. (Etymologically, the word *person* means "to sound through"; coming from the Greek *persona*, it refers to a dramatic mask with a speaking hole.) According to Lacan, the jubilation that creates the essentially false and paranoid ego in the infant

occurs when its gaze confronts the image of a fully contoured and coordinated body at the very time (six to eighteen months) it is beleaguered by
a motor incapacity that renders it more helpless and defenseless than perhaps any other mammal of the same age.[7] The intense motor incapacity
and uncoordination, resulting from "prematuration" (or, in evolutionary
terms, from neoteny), engulfs the infant in an almost cinematographic
world of uncontrollable visions. One of these mysticlike visions is of itself
(or the mother) with a coordination and in a place where it does not in
fact exist, along the rectilinear mirror plane. This form of mystical iden-
tificatory representation with an image or imago Lacan designates as
"image-inary." As a fictional form of the I, it is comforting and effects the
discrete sense of self from toddler on into adulthood, the sense of self that
has been catered to by American ego psychology in contradistinction to
the original Freudian insights and painstaking deconstructions of a psyche
(psycho-analysis) formerly presumed to be whole. The Lacanian psychoanalytic revamping of the myth of Narcissus suggests that what we perceive to be our body, as the locus of our "self," is in fact plastic,
malleable; and indeed, the lability of the imaginary view of self has come
to the fore in the first technology-mediated glimpses of a new image of the
human body: Earth from space.[8] This rapidly proliferating image, now
recognized as our ecological or biospheric home, will, with further population growth, interspecies interdependencies, and optimization of global
media, begin to be re-cognized as body.[9]

Already the shift from biosphere-as-home to biosphere-as-body has
become apparent in the scientific work of James E. Lovelock, whose Gaia
hypothesis, with mythical allusions of its own, has inspired a planetary
search for "geophysiological" climatological and biogeochemical mechanisms. Biospheric individuality was already recognized by Julian Huxley,[10]
who wrote:

> [T]he whole organic world constitutes a single great indi-
> vidual, vague and badly co-ordinated it is true, but none the
> less a continuing whole with inter-dependent parts: if some
> accident were to remove all the green plants, or all the bac-
> teria, the rest of life would be unable to exist. This individ-
> uality, however, is an extremely imperfect one—the internal
> harmony and the subordination of the parts to the whole is
> almost infinitely less than in the body of a metazoan, and is
> thus very wasteful; instead of one part distributing its sur-

plus among the other parts and living peaceably itself on what is left, the transference of food from one unit to another is usually attended with the total or partial destruction of one of its units.

As positivists, materialists, or physical reductionists in the Western scientific tradition, we would like to think that the picture of the body as an adequately closed topological surface is necessary and sufficient prima facie self-evidence—for the self. And so it is within a certain rectilinear closure. However, as we—and even this coauthorial "we" must be put in quotation marks as we ponder the self, the subject, the person, et cetera—intimated, the egotistic I is clear only in the sense of a fundamentally fictional or topologically displaced mirror image; there is nothing behind the mirror. Emphasizing tactility rather than vision, on a sensual level it is easy to imagine a conception of the human environment as beginning with the fingernails, hair, bones, and other substances no longer considered to be body parts because they are bereft of sensation. Conversely, technological introjection exemplified by devices such as tele-vision (video, movies, et cetera) and tele-portation (automobiles, airplanes, and so forth) suggests a topological extension of the human into what formerly would have been considered the environment. Therefore the body, the material or corporeal basis for "self," has no absolute time-independent skin-encapsulated topological fixity. It is a sociolinguistic psychoanalytic evolutionary construct. Mucus, excrement, urine, spittle, corpses, pornography, and other detachments from and marginal representations of the human body call its essential hegemony, its universal nature, into question.

Chastising the Spanish artist for painting unrepresentative cubistic abstractions, a layman withdrew a photograph of his wife from his pocket and held it up to Picasso with the admonition, "Why can't you paint realistically, like that?" "Is that what your wife really looks like?" Picasso asked. "Yes," replied the man. "Well, she's very small, and quite flat." Our working assumption of what the self is—like the layman's view of what his wife "really looks like"—is based on a model of representation that takes far too much for granted. Representation itself has, in postmodernist philosophy, fallen into disfavor in a manner similar, perhaps, to that in which figurative realistic painting fell into disfavor with the innovation of the camera. This does not mean that the possibilities of representational or propositional truth, of the correspondence theory of reality still so entrenched in science, are necessarily dead; on the other hand, the difficulties posed by the evidence of quantum mechanics, not least of which is the philosophical nonsolution of the Copenhagen interpretation of the structure of the atom, suggest that most scientific models of reality may be neither so enlightened nor au courant as they assume. Indeed, what is in question is the very possibility of modeling reality at all.

Psychoanalytically, when we broach the topic of castration, amputation, dismemberment, the infant's polymorphic perverse sensations and perceptions of the body being, as in a picture by Hieronymus Bosch, in bits and pieces are probably close to the true state of nature, if such a state there be. In other words, the infant's primordial presocialization experience of the world should not be considered inaccurate but rather, precisely because it precedes sociocultural linguistic norms, less prejudiced and potentially more "realistic." And, apart from parturition, there may be a biological basis for these perceptions, which, later in life, are recalled as amputation, castration, dismemberment. Permitting ourselves a wee bit of abstraction here we splice in a couple more comments by Huxley (1912):

> . . . certain bits of organic machinery are of such a nature that it is physically impossible for the animal to live at all if they are seriously tampered with. It is just because our blood-circulation is so swift and efficient and our nervous system so splendidly centralized that damage to heart or brain means almost instant death to us, while a brainless frog will live for long, and a heart-less part of a worm not only will live but regenerate. Thus here again sacrifice is at the root . . . and only by surrendering its powers of regeneration and reconstitution has life been able to achieve high individualities with the materials allotted her. . . . We have seen the totality of living things as a continuous slowly-advancing sheet of protoplasm out of which nature has been ceaselessly trying to carve systems complete and harmonious in themselves, isolable from all other things, and independent. But she has never been completely successful: the systems are never quite cut off, for each must take its origin in one or more pieces of previous system; they are never completely harmonious.

Given the abiding prevalence of an imaginary or representational worldview in Western science, it is impossible to overestimate the theoretical importance of this relatively abstract, nonrepresentational splicing or grafting that crosses cellular, species, and taxonomic boundaries. Light, no less than matter, cannot be understood simply as a collection of particles but must also be comprehended as a wave: with quantum mechanics the Democritean atomistic Newtonian worldview has come to a functional end, although the momentum of scientific discourse has prevented it from reckoning with the consequences of this theoretical shipwreck.

Comparable with the end of the Newtonian age in physics, evidence of the dwindling of an atomistic model of organismic identity in the biological realm is reflected by the debate over the essential unit of selection in Darwinian evolution: whether it is really the gene (inside the organism), the "individual" competing organism (as Darwin stressed), or group levels such as species or multicellular assemblages. Hierarchy theory entertains species and multicellular assemblages—extended phenotypes outside the organism and beyond the traditional confines of the self—to be the crucial units of selection. The paradoxical notion of group selection, against which neoDarwinists seem to rant,[11] is patently necessary to explain epochal evolutionary transformations such as those from protoctista colonies to the first plants, animals, and fungi.

The minimal autopoietic, or living, system is the membrane-bounded cell. A cell, or any other autopoietic entity of even more complexity, undergoes continual chemical and energetic transformations easily recognizable as "being alive." In the process of this ubiquitous metabolism, each living

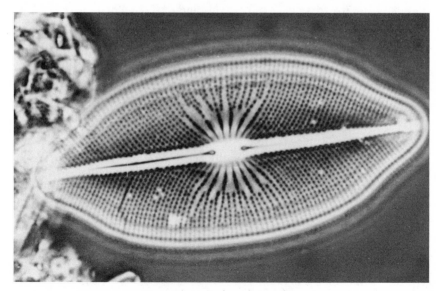

Figure 3.1 *Navicula*, a diatom alga.

entity is materially contained within at least a membranous boundary of its own making. In addition to the universal plasma membrane of all living cells, other boundaries, for example, cuticles, skin, thecae, or valves ("shells," "tests," equivalent terms in diatoms [figure 3.1]), are self-produced. Such borders include the black-haired, bumpy skin of humpback whales, the glycocalyx of some amoebas, the hard overwintering thecal coat of hydra eggs, and the waxy cuticle of a cactus. Minimally the autopoietic unit produces the plasma membrane, but often beyond the lipid-protein membrane are the cellulosic walls, coccolith buttons, or siliceous spines—complex material extensions found just outside, adjacent to, or attached to the universal membrane. Live beings continually construct, adjust, and reconstruct their dynamic structures by which they are bounded.

We recognize autopoietic entities as "individuals" or "individual organisms." A tree, a potted plant, a swimming euglena, and a cat are immediately perceived as single living organisms. Minimally, all such autopoietic entities—cells—are comprised of at least one genome: a DNA-containing nucleus or nucleoid (that is, the total of all the organism's genes) and the internal cellular apparatus (RNA-driven, protein-synthetic, and ribosome-studded) of that genome.

What is the lowest common denominator of individual life? The minimal entity, a single genomic system, is the bacterial cell. Most bacteria are metabolizing units of some two to five thousand genes and their proteins bounded by the dynamic cell membrane. Multicellular bacteria, for example, *Polyangium*, *Fischerella*, and *Arthromitus*—there are myriads of them—are comprised of many copies of the same genomic system. Filaments, tree-shaped branches or gelatinous spheres, multicellular bacteria are composed of homologous genomic systems in direct contact with one another (figure 3.2). In some cases, like swarms of cyst-forming myxobacteria (for example, *Stigmatella*, *Chondromyces*, or *Myxococcus* [figure 3.3]), the component cells sense each other and fuse to form larger structures in which no membranes are breached. In others, as when the akinetes (filamentous propagules) of a cyanobacterium float away, the genomic systems disperse. Here two examples of multicellular bacteria, *Stigmatella* and *Arthromitus*, are shown.

All organisms of greater morphological complexity than bacteria, that is, nucleated or eukaryotic organisms (whether single-celled or multicellular), are also polygenomic. They have selves of multiple origins. These "selves" are comprised of heterologous different-sourced genomic systems that each evolved from more than one kind of ancestor.

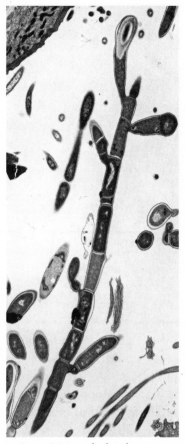

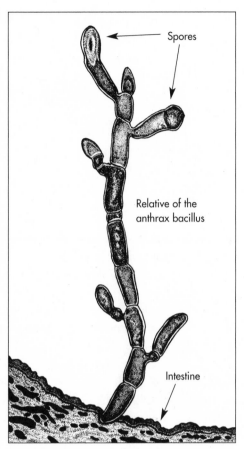

Figure 3.2 Branched *Arthromitus*.
Electron micrograph by David Chase.

Figure 3.3 Branched *Arthromitus*.
Drawing by Christie Lyons.

We now see a possible correspondence of the "sense-of-self" to "autopoietic entity" or "live individual." All individuals, all living organisms, actively self-maintain. From the early Archean eon (3,500 million years ago) and its bacterial inhabitants through the protoctists of the Proterozoic eon (2,500 to 541 million years ago) and the fungi, plants, and animals of the Phanerozoic eon (541 million years ago to the present), the "sense-of-self" seems synonymous with the nature of autopoiesis. Boundaries resist breaching while biochemistry acts to maintain integrity. Life's nature is to interact with the material world to incessantly integrate its components, rejecting, sorting, and discriminating among potential food, waste, or energy sources in ways that maintain the integrity of the organism.

What is remarkable is the tendency of autopoietic entities to interact

Figure 3.4 *Stigmatella,* a multicellular bacterium. Drawing by Christie Lyons.

with other recognizable autopoietic entities. These interactions may be neutral, as in an amoeba and a pebble; that is, no obvious reaction may occur at all. Two approaching organisms may be indifferent. Alternatively, two heterologous organisms may be destructive—disintegrative—toward each other. One, for example, may produce extracellular enzymes that destroy the other and, relieving it of its autopoiesis, break it down to component metabolic parts. The resulting chemical breakdown products may then be used as food. This may be called a trophic relation whereby the still-intact autopoietic being, the feeder, consumes and incorporates the chemical components of its victim food.

Though relations between organisms may be disintegrative or neutral, interactions between autopoietic entities that lead beyond destruction to integrative mergers we find most fascinating. Such mergers (such as fertilization or partner integration in symbiosis) lead to autopoietic entities of still greater complexity. For example, the integration of a fungus that attacks an alga for nutrients often—perhaps twenty-five thousand times— has led to a balance between the disintegrative responses of both fungal and algal partners. Eventually a lichen emerges. A lichen is neither a fungus nor an alga—as a "lichen" it is a composite symbiotic complex that itself is a "self" at a more complex level of organization than either fungus or alga.

With regard to the latter-day three-dimensional pointillist elaboration of the arcane immunity of virus-infected bacteria, we are admonished to

ponder the connections. The AIDS-infected human differs little—in principle—from the *E. coli* bacterium infected with lysogenic bacteriophage.
The "independence" of the nervous system (mind) from the immune
system (body) is severely questioned. Candace Pert defiantly speaks only
of bodymind or mindbody.[12] Interviewed by her friend Nancy Griffiths-
Marriott, she points to an overemphasis of the blood-brain barrier and
the model of the nervous system as a network of penetrating, penile-
shaped cells that control the body. Pert emphasizes that monocytes cross
that "barrier" within seconds; furthermore, these cells of the immune
system transform to become the glial cells of the nervous system. (Glial
cells are ten times more abundant than neurons in the vertebrate nervous
system.) Like gut and brain cells, such monocytes bear neuropeptide
receptors—surface proteins—sensitive to the endorphin peptides—natural or endogenous drugs inside the individual—of the neuroimmune
system that bring on feelings of elation and ecstacy. Neuropeptides, small
communicative molecules, include vasointestinal peptides and endorphins
that signal to monocytes. Such proteinlike molecules attach to the cell
receptors at the surface of gut or brain or monocyte cells at the same place
the AIDS virus gets stuck. No, says Pert, there is no mind/body, con-
troller/controlled, male/female, neuron/glial cell dichotomy. Rather there
is "mindbody-bodymind," a dynamic system kept informed by devastating news, transforming monocytes, neuropeptide messengers, and hundreds of other integrating mechanisms that confirm the mobile self.

Beginning as latter-day evolution of bounded endosymbiotic bacterial
communities we—as densely packed biomineralizing complexes of eukaryotic cells—should not be too sanguine about the longevity of the modern
notion of self. Already in the nineteenth century Samuel Butler clearly and
successfully deconstructed personality by parasitizing Charles Darwin's
texts. Between the human ovum and the octogenarian, held Butler, lie differences greater than those between us and other species. What with the
vagaries of memory and experience, it is essentially arbitrary to believe that
the zygote and the eighty-year-old are the same person, whereas the father
and the son have different selves. Genotypically we may argue with Butler,
but to do so phenotypically would be a far more difficult chore. Butler
demonstrates the essential arbitrariness of our definitions of organismic
identity, of organic integrity and "individuality," even more strikingly by
taking the case of a moth. Here we have a being, Butler says, that undergoes radical bodily change between egg and chrysalis, between pupa and
winged insect; and yet the only time we say it dies is after the adult moth

form stops moving its wings, despite the other radical phenotypic changes during which the genotype has nonetheless been preserved. We might as easily, Butler reminds us, have chosen to consider the transfer from egg to chrysalis or from chrysalis to moth as "death"—and construed the demobilization of the moth as a sloughing-off similar to the shedding of a skin. Indeed, to seriously consider death at all entails a certain ignorance—a certain disregard for the continuity of the "personality" despite its radical transformations. So you see that with this figure in which the moth's "self" is held aloft on the tenterhooks of quotation marks "we" have provisionalized identity—not least of all by avoiding the traditional figure of the rectangle that enframes the essay, representing thoughts in an enclosed form that seems to mirror the hegemony of a rigidly structured Platonic body. Topologically the self has no homuncular inner self but comes

Chapter 3 Notes

1. Maturana and Varela, 1973.
2. Fleck, 1979.
3. Habermas, 1987.
4. Leenhardt, 1979.
5. Snell, 1960.
6. Elias, 1978.
7. Lacan, 1977.
8. Sagan, D. 1990b.
9. Sagan, D. 1990a.
10. Huxley, J. 1912.
11. Dawkins, 1976, 1982.
12. Pert and Griffiths-Marriott, 1988.

part two
chimera

Chimera and his siblings Hydra and Cerberus were offspring of Echidna, the monstrous serpent woman born of Earth who became pregnant when she mated with her brother. We choose these miscreants because their mixed features and strange powers are reminiscent of the different kinds of bacteria that fused to form the ancestors of all visible life-forms, from seaweeds to humans to whales. The fire-breathing Chimera has the foreparts of a lion, the midsection of a goat, and the behind of a serpent. Cerberus, reverse bouncer of the underworld Hades, who prevented people from leaving, is a three-headed dog with the tail of a serpent. The Hydra not only is a nine-headed serpent but also, in a sort of mythological parable for the raw power of reproducing life, grows back two heads for each one cut off.

Ironically, the term hydra also refers to marine animals with a cylindrical body and a ring of tentacles surrounding the mouth that live attached to rocks or plants but can also detach and float in the water; these sea creatures, born pale, sometimes merge with green algae, becoming functionally photosynthetic. In Greek mythology the Hydra is slain by Hercules, who devised the expedient of cauterizing the neck immediately after severing each of the nine heads; this prevented them from growing back doubly. The Chimera, for its part, was slain by the Corinthian hero Bellerophon with help from Pegasus, another pre-biotech hybrid combining the body of a horse with the wings of a giant bird. Cerberus barked viciously at Hercules, and his spittle gave rise to a poisonous plant named aconite or hecateis or, in modern times, as wolfbane. Wolfbane, a key ingredient in the magical ointment empowered Thessalian witches with the ability to fly; Medea tried to poison Theseus with it.

As you will see, our genetically blended ancestors are the real-life counterparts to the monsters of classical Greek imagination. They remain unslain. In fact, as mitochondria inside the muscles of the hero holding the sword, they are part of the slayer.

The detailed way in which all live organisms are complex with numerous interacting components is the central concept of this set of nine essays (chapters 4 through 12). Integration of microbial communities and human technological extensions, or "the self," requires energy and matter fluxes, as it has for thousands of millions of years, since the origin of life. Humans are one kind of self, but we are composed of smaller selves, and we form parts of the more inclusive selves.

Power to the Protoctists

LYNN MARGULIS

Some 250,000 species of "water neithers," extant organisms that are neither plant nor animal, tell us about life, sex, and nature. Most of these "beasties," as he called them, have been disenfranchised or demeaned since their existence was discovered by Antoni van Leeuwenhoek in the seventeenth century.

TWO OLD KINGDOMS

Perhaps because our brains are divided into two halves, our bodies into two sexes, and our language into two genders, the human tendency to dichotomize—to divide things into either *this* or *that*—is almost irresistibly strong (see chapter 20). According to traditional systems of classification, anything alive must be either plant or animal. But taxonomy, or the placing of organisms into categories, is not just an educational exercise—it colors our values, affects our actions, and embeds hypotheses that promote mind-sets that pervade our thinking. Furthermore, such mind-sets may persist even when the classification system becomes obsolete. So it is with the plant/animal legacy. If we view microbes, all those organisms invisible to the unaided eye, as mere "germs," hence unworthy of our consideration, we slight those organisms that provide our air and fertilize our soil, and we separate essential processes from the web of life. We codify our ignorance and preclude learning to use the recycling and gas production skills of the so-called lower beings. The old labels impede the spread of knowledge about the mutually dependent diversity of life and its importance to our well-being.

The two-kingdom system—and our position in it—started unraveling with the invention of the microscope in the late 1600s, which enabled the Dutchman Antoni van Leeuwenhoek for the first time to see subvisible organisms. Those that swam reminded him of tiny animals, so he named them animalcules. Microscopic beings that didn't move or were green he called tiny plants. But, on closer examination, none of the vast world of microorganisms is so easily pigeonholed.

What about an organism such as *Euglena gracilis*? With a microscope, one can see in *Euglena* green parts that look just like those in the leaves of a plant. Because it photosynthesizes, *Euglena* would seem to be a vegetable. But *Euglena* cells also swim. Each has a single moving appendage closely resembling a human sperm tail. Swimming, a kind of locomotion, traditionally is a defining trait of animals. Botanists claimed *Euglena* was a plant, zoologists classified it as an animal, and potential biology students fled to study more logical fields, such as chemistry.

As observations of the microcosm blossomed, more and more oddballs appeared that further muddied the distinctions between plant and animal: Are malarial parasites animals? Are slime molds not fungi and therefore plants? Aren't diatoms phytoplankton, hence marine plants? Are amoebae single-celled animals? And is dry yeast dead? Or is it an animal, a fungus, or a plant?

The problem lies not with the swimming green *Euglena* but with our category errors: the flawed classification system that promotes dogmatic ignorance. The two-kingdom system—formalized in the eighteenth century by Carl von Linné (Carolus Linnaeus)—developed in a hostile world. Floods, earthquakes, plagues, and pestilences, which humans could neither understand nor master, seemed to have nothing to do with living nature. Little wonder that our ancestors comforted themselves with the belief that they were the apex of God's creation, given dominion over nature and set apart from it as unique and independent. Western science then embraced humans' egocentric view of themselves as the pinnacle of a linear evolution from the lower, "primitive" to the highest, "most evolved" forms, us.

The combination of new powerful microscopes, molecular biology, and modern genetics and paleontology has enabled refinements of taxonomic distinctions to the level of genes and proteins. These sophisticated methods upset old biological dichotomies. It is indisputable that all life on Earth today derived from common ancestors; the first to evolve—and the last to be studied in detail—were tiny, oxygen-eschewing bacteria. So significant are bacteria and their evolution that the fundamental division in life-forms is not that between plants and animals but that between prokaryotes (bacteria, composed of small cells with no nuclear membrane surrounding their genes) and eukaryotes (all other life-forms, including humans, composed of cells with those nuclear membranes). In the first two billion years of life on Earth, bacteria—the only inhabitants—continuously transformed the planet's surface and atmosphere and invented all of life's essential, miniaturized chemical systems. Their ancient biotechnology led to fermentation, photosynthesis, oxygen breathing, and the fixing of atmospheric nitrogen into proteins. It also led to worldwide crises of bacterial population expansion, starvation, and pollution long before the dawn of larger forms of life.

Bacteria survived these crises because of special abilities that eukaryotes lack and that add whole new dimensions to the dynamics of evolution. First, bacteria routinely transfer their genes to other bacteria very different from themselves. A recipient bacterium can use the visiting, accessory DNA (the cell's genetic material) to perform functions that its own genes cannot mandate. Bacteria exchange genes quickly and reversibly, in part because they live in densely populated communities. Consequently, unlike other life, all the world's bacteria have access to a single gene pool and hence to the adaptive mechanisms of the entire bacterial kingdom. This extreme genetic fluidity makes the concept of species of bacteria meaningless. The result is a planet made fertile and inhabitable for larger life-forms by a worldwide system of communicating, gene-exchanging bacteria.

Bacteria also have a remarkable capacity to merge transiently or permanently with larger organisms. Alliances may dissolve or become permanent. Fully ten percent of our own dry weight consists of bacteria, some of which—such as those microorganisms in our intestines that produce vitamin B_{12}—we cannot live without. Mitochondria live inside our cells but reproduce at different times using different methods from the rest of the host cell. They are descendants of ancient, oxygen-using bacteria that were either engulfed as prey or invaded as predators. These bacteria took up residence inside ancient motile cells to form an uneasy alliance. Waste disposal and oxygen-derived energy were bartered for food and shelter. Without healthy mitochondria, the plant or animal cell cannot breathe. Therefore it dies.

Such symbiogenesis, the merging of organisms into new collectives, is a major source of evolutionary change on Earth. Early first mergers led to the evolution of protoctist cells, or eukaryotic microorganisms. Protoctists, our most recent, most important—and most ignored—microbial ancestors, are woefully misunderstood, even by scientists. Protoctists invented our phagocytosis, or ingestive animal sort of digestion. They originated cell movement and visual and other sensory systems. Speciation, cannibalism, genes organized on chromosomes, and the ability to make hard parts (like teeth and skeletons) all began with protoctists. These complex microscopic beings and their descendants even developed the first male and female genders and our kind of cell-fusing sexuality-fertilization involving penetration of eggs by sperm.

Scientists thus have discovered that bacteria not only are the building blocks of life but also occupy and are indispensable to every other living being on Earth. Without them, we would have no air to breathe, no nitrogen in our food, no soil on which to grow crops. Without microbes, life's essential processes would quickly grind to a halt, and Earth would be as barren as Venus and Mars. Far from leaving microorganisms behind on an evolutionary ladder, we are both surrounded by them and composed of them. The new knowledge of biology, moreover, alters our view of evolution as a chronic, bloody competition among individuals and species. Life took over the globe not by combat but by networking. Lifeforms multiplied and grew more complex by co-opting others, not just by killing them.

Discovering the microcosm within and about us changes—indeed, reverses—the way we look at living things and picture their evolution on the planet. Because all life on Earth evolved from bacteria, it makes more sense now to think of beetles, rosebushes, and baboons as communities of bacteria than it does to think of bacteria as tiny animals or plants. This new worldview, in turn, requires a new, more representative labeling system.

But ignorance and resistance have stalled that process. Overhauling the two-kingdom convention—a vast information retrieval system on which so many depend—would require, for starters, changing how we file and compile bibliographies, how we handle agricultural permits and customs declarations, and how we compute ocean diversity and measure ecological stability. More important, the traditional two-kingdom system and the attitude it embodies endure because shifting from the belief in "man, the highest animal" to a more egalitarian view of the world that respects and empowers all life is too drastic a mental move. To admit that our ancestors are bacteria is humbling. It has disturbing implications. Besides impugning human sovereignty over the rest of nature, it challenges our assumptions of individuality, uniqueness, and independence. It even violates our view of ourselves as discrete physical beings separate from the rest of nature and—still more unsettling—questions the alleged uniqueness of human intelligent consciousness.

Not surprisingly, the idea of the subvisible microcosm in strong, continuous, and consistent interaction with us still has not expanded to the world of the ordinary. Only in the 1970s did any scientists begin to take seriously alternatives to a two-kingdom system. And now some want a three-domain system that privileges the prokaryotes. Most of humanity—including those who make political decisions about life's diversity—still clings to a two-kingdom view. Our culture discounts the importance of beings that are neither pet, nor relative, nor food, nor directly useful to us—especially any we can't even see. Many ignore the microbially based productivity and waste-recycling capacity of wetlands, for instance, because wetlands seem useless as real estate if not drained.

NEW ALLIANCES

Thoughtful biologists, those of us who work with live organisms, have shifted to new taxonomies. We recognize that—from multicellular bacteria to marmosets—all living forms are coevolving products of nearly four billion years of evolution. None is "more evolved" than any other. Our scheme[1] reorganizes life into two superkingdoms: Prokarya and Eukarya, which together comprise five kingdoms, each with their subkingdoms. The kingdoms are listed here in order of their origin:

Bacteria | Protoctista | Animalia | Fungi | Plantae

In this categorization, the perhaps half-million different kinds of bacteria, fungi, and protoctists that are neither animals nor plants, and such former "misfits" as slime molds, yeasts, and *Euglena*, have finally found a niche. (Because viruses are incapable of any metabolic transformation, including DNA replication outside a living cell, they are not alive and are not members of any of the five kingdoms. Unrelated to each other, they are probably runaway fragments of diverse origin.)

Bacteria—the most metabolically diverse and smallest cells on the planet—reproduce primarily by direct cell division. Without chromosomes, unlike larger life, bacteria have a more informal arrangement of DNA that probably allows more flexible and frequent gene exchange between them. The absence of countable protein-rich chromosomes probably precludes the rigid relation between sex and reproduction frequent in many animals and plants. The capabilities of bacteria include the digestion of cellulose in the guts of cows, the coloring of swimming pools blue-green with photosynthetic cyanobacteria, and the "fixing" of nitrogen,

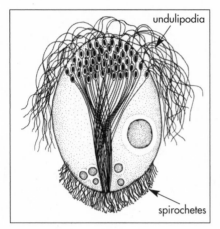

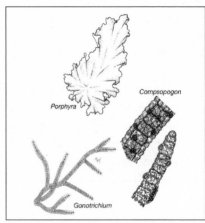

Figure 4.1 *Stephanonympha* sp., an archaeprotist (no mitochondria). Drawing by Sheila Manion-Artz.

Figure 4.2 Red algae, *Rhodophyta.* Drawing by Sheila Manion-Artz.

converting it from an inert gas of the air to a usable form in water and soil. Every spoonful of garden soil contains some ten·billion bacteria. The total number in any person's mouth is comparable to the number of people who have ever lived. We rely on our personal bacterial populations to help us digest our food and to keep us healthy by restraining the overgrowth of unwelcome microbes. To be kept alive, babies born without their microbial symbionts require "germ-free bubble" lodging at the cost of $100,000 per day!

All the nucleated organisms, that is, eukaryotes that are neither animals, plants, nor fungi, are protoctists.[2] This huge group includes ciliates, amoebae, the malarial parasites, *Giardia*, slime molds, photoplankton, the seaweeds, and single-celled photosynthetic swimming microbes such as *Euglena*. Protoctists are aquatic: some live primarily in the oceans, some primarily in freshwater, some in the watery tissues of other organisms. Some are parasitic. Nearly every animal, fungus, and plant—perhaps every species—has protoctist associates. While most are harmless, certain protoctists cause tropical diseases (for example, Chagas' disease, giardiasis, malaria, and sleeping sickness) or red tides. Because nearly all photoplankton are protoctists, this group also forms the basis of the ocean food chain. Protoctists show remarkable variation in cell organization, cell division, nutrition, and life cycles but are far less metabolically diverse than the bacteria (figure 4.2).[3]

The fungi kingdom includes all the yeasts and most of the molds, as well as truffles, puffballs, and mushrooms. These live mostly on land.

After the bacteria, fungi were among the first to leave strictly aquatic habitats. Fungi are tenacious microbes. Most are able to resist desiccation. Some grow in acid; others survive in environments largely lacking nitrogen, an essential ingredient for all life. Certain fungi are responsible for rotting fruit, raising and molding bread, smelly feet, ripening cheese, fermenting beer and wine, and producing antibiotics such as penicillin.

Those who speak only for the special interests of human beings fail to see how interdependent life on Earth really is. We cannot view evolutionary history in a balanced manner if we think of it only as a four-billion-year preparation for us "higher" animals. Life's history has been mainly microbial—from archaea (archaebacteria) in submarine boiling methane-rich springs to fungi that sprout their spores through the heads of ants. We are recombinations of the metabolic processes of bacteria that appeared during the accumulation of atmospheric oxygen some two thousand million years ago. We tend to separate ourselves from the rest of life, yet without the others, especially the microbial others, we would sink in our feces, drown in our urine, and choke on the carbon dioxide we exhale. Like rats, we have done well distinguishing ourselves from and exploiting other forms of life. These delusions cannot last.

Chapter 4 Notes
1. Margulis and Chapman, 2008.
2. Margulis et al. 1990.
3. Margulis et al. 1993.

Prejudice and Bacterial Consciousness

LYNN MARGULIS

The more I learn about bacteria, the more astounded I become.

I never heard the term "African American" until I was already a mother of four. My classmates and neighbors, many of them great-grandchildren of former slaves from Texas and Mississippi, were called "niggers," "schwartze," "nigros," or "colored people" when I was a child. In those days, my strongest emotion was the impetus to escape from those who unselfconsciously used such appellations on Chicago's South Side. Today, although prejudice may prevail in privacy, such repulsive names for so many of our people have been silenced.

A wave of the old emotion that led me to despise those speakers, to escape that society of bias, comes over me today when I overhear people dismiss bacteria in comparable ignorance. They label bacteria "germs," disdaining them as enemies and bragging of modern medical victories.

Little fuels my misanthropy more than reading that pharmaceutical companies will rid the world of the anthrax bacterium once and for all, or seeing labels on antibacterial soaps and claims for over-the-counter antibiotic medicines that guarantee to vanquish the deadly germs of children's hands and ears. My anger bubbles up when health-food stores advertise *Spirulina*, the green food supplement, as an alga. I was dismayed when my radio heroine Laurie Sanders, wonderful host of the program *Field Notes* on WFCR-FM, our local radio station, who knows better, pandered to public misconception. Even she implied *Spirulina* is a type of seaweed. In fact, *Spirulina* are edible, healthful bacteria. I nearly became apoplectic, but what's the fuss? Why shouldn't African Americans be called "niggers" and bacteria be called "germs" when, after all, that's what they are, aren't they?

No! As Lewis Thomas (1913–1993), probably the most talented and intellectual physician in the history of this country, author of *The Lives of a Cell* and *Late Night Thoughts On Listening to Mahler's Ninth*

Symphony, noted, disease bacteria—like armed bandits, not tender fathers—are infrequent freaks of the subvisible world. Most bacteria have far more important things to do on this Earth than to devour our tissues while we are still alive, drink our blood when we are old and weak, or fight with us over who will eat our food first.

Indeed, we have much to learn from bacteria—including how to prepare soil for plants, recycle nitrogen, and conserve water—and about bacteria.[1] Bacteria invented photosynthesis and swimming, evolving prior to any animals or plants. Some bacteria generate energy in the dark from fool's gold (pyrite). Others make rocks. Our cultural prejudices, our haughty deprecation, prevent access to their ancient wisdom and the salutary effects of dialogue.

I might rage on. More bacteria inhabit your intestinal tract right now than the number of people who lived on Earth in the last million years. Your body contains a greater number of bacterial than human cells. Some bacteria with tiny magnets in their bodies orient and swim north more accurately than fish. Some bacterial cells promptly begin to reproduce when, after forty-eight hours of roiling around in boiling water without pause, the water cools. Still other bacteria can eat out the intractable (for us) four percent protein between carbonate crystals of a clean clamshell.

Those who hate and want to kill bacteria indulge in self-hatred. Our ultimate ancestors, yours and mine, descended from this group of beings. Not only are bacteria our ancestors, but also, if I am correct, as the evolutionary antecedent of the nervous system, they invented consciousness. The effects we recognize as sensitivity to light, sense of touch, hearing, smell, and indeed our senses in general evolved from a property properly called "bacterial consciousness."

The cells that compose animal and plant bodies tend to be larger and more complex than those of bacteria. The chemistry and fossil history of bacteria, though, convince everyone that they must have been our ancestors. Bacterial genes, like ours, are made of DNA. Bacterial cell membranes are composed of fatty compounds linked to phosphoric acid and studded with proteins; so are ours. No scientist, science teacher, or even reader of science fiction doubts for a minute that all other organisms, including mammals like us, evolved from bacterial ancestors.

For years, with Dr. Michael Dolan and Dr. Dennis Searcy, both professors at the University of Massachusetts at Amherst, and other colleagues (Drs. Ricardo Guerrero and John Hall) I have been working on the evolutionary transition from bacterial cells to those of algae, animals, and plants: cells with nuclei.[2] No fellow scientists, no matter how vehemently they dispute the details, can disprove our central tenet: the nucleated cell of all protoctists, fungi, animals, and plants evolved from a community of bacteria by way of merger, literal incorporation of one kind of bacterium by a different bacterial type.

The larger bacterium lacked a cell wall and incorporated smaller others, probably after a long fight. The larger one, an archaean (=archaebacterium), belonged to this newly discovered bacterial "domain."[3] Clues reveal that animal and plant cells still have the chemical capability to make noxious hydrogen sulfide (H_2S gas). This virtuosity, typical of today's archaebacteria (like the *Thermoplasma acidophila* studied by Searcy), we think is an ancient legacy.

The first of our tiny ancestors, the archaebacterial partner, fermented sugar for food and energy the way our cells still do. Hydrogen sulfide, the gas of rotten eggs, quickly reacts with oxygen gas, and from the beginning H_2S production by live cells probably detoxified the increasing quantities of ambient oxygen.

The once independent two types of bacteria (both the smaller oxygen-breather and the larger wall-less H_2S-forming archaebacterium in which it resided) were already "conscious" entities. A dictionary definition of *consciousness* is "awareness of the world around one." Evidence for bacterial awareness abounds in the scientific literature. Many bacteria glide toward oxygen gas and away from sulfide or swim to edible sugars and away from strong acids or dangerously high salt solutions. Others eschew oxygen or cold water but make a beeline for the seaside mud hole where H_2S bubbles out. Many bacteria respond to light by basking in it. When light intensity is too high, some synthesize "sunglasses," brown pigment that prevents sunburn. Others sense desiccation. They dry out entirely even while bathed in water! Such strategies lead them to overwinter, hibernate, or estivate, in dry mud.

I continue to work on an idea, ignored or rejected by many professionals, about the origin of nervous system sensitivities, which is that the projections found on the tips of our sensory cells evolved from mergers with a swimming bacterium.

The general principle is easy to grasp. We know it to be true of Australian flightless birds, the Ozark blind salamander, and the Mexican eyeless cavefish. Wings, eyes, tails, flowers: no important ancestral trait has ever been lost without leaving a trace. If in evolutionary history a once useful feature is selected against (that is, not perpetrated in a lineage from parent to offspring), the feature never arbitrarily vanishes. Rather, some vestige—wing bones, eye sockets, minuscule petals—remains.

Details of cells or of organisms made of cells help trace evolutionary history. In mammals, the cells of the tongue's taste buds, the inner-ear cells required for hearing, and light-sensitive cells in the retina of the eye have peculiar features in common. Even cells of the semicircular canal (our balance organ, the stimulus receiver that tells us whether we stand on our feet or upside down on our head) share a detailed structure that provides a clue to their origin. The notable feature shared by the projections at the ends of these sensory cells is a distinctive, weirdly symmetrical arrangement of tubules that proffers an indication of common ancestry. The human brain, like all brains and nerve cells in general, is replete with these skinny, often very long tubules. Those inside nerve cells are called neurotubules.

My University of Massachusetts and Barcelona colleagues and I are testing the idea now. The distinctive skinny tubules in the nerve and other cells of animals and plants, by our reckoning, originated from similar tubules in once–independent swimming bacteria. The skinny tubules of sperm tails and sense-organ cells are easiest to trace because of their invariant pattern: nine tight pairs of tubules surround a looser pair in the middle.[3]

I think these [9(2)+2] patterns evolved in once free-swimming bacteria that were acquired as symbionts. Fusions of three different kinds of once independent bacteria led to sensitive, swimming, sugar-eating, oxygen-breathing, wall-less H_2S-forming larger-than-bacterial-cells nucleated cells. Nucleated common ancestors proliferated all over the world and left fossils a thousand million years ago, long before *any* animals swam the sea or roamed the plains.

From the beginning these sensitive bacteria corkscrewed their way through microbial mats, decaying bodies in tropical, oxygen-depleted mud, as we see many do today in moist, rich humus or animal tissue. The descendants of these bacteria are likely still to thrive in such places whose essential features persist, such as seaside muddy shallows or the hindguts of termites depleted in oxygen gas. The general principle is well understood: After a successful style of life evolves (such as the ability to form spores that survive boiling water or photosynthesis that uses H_2S gas but gives off no oxygen), that lifestyle is retained as long as the specific environment persists. (Think of hot springs or sunlit, muddy shorelines.)

From the directed behavior of many different kinds of thriving bacteria in such habitats, we infer a sort of "microbial mind," healthy choices of where to go, how long to stay, with whom to congregate, when to leave,

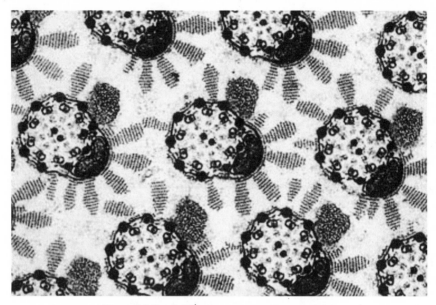

Figure 5.1 Sperm tails. Electron micrograph by Bjorn Afzelius.

how to find a mate, with whom to to ensure descendants. Indeed, bacteria in their natural environments, rather than in captivity in laboratories, all live in well-structured communities. Prodigious in growth rates, they need each other. As one type excretes an acid or a sugar, these waste compounds become food that specifically attracts others.

The vast numbers of incessantly moving but mute bacterial denizens ignore us as they eat, grow, and reproduce, as we ignore them. Very few, only the "freaks," poison us or directly feed on us in ways that injure us. It is the notorious self-centeredness of our species that leads to the manic claim that all bacteria are killer germs to be eradicated from our lives.

In the merger process among bacteria community members, relationships changed: aggression gave way to truce, accommodation followed cannibalism and predation, and cohabitation succeeded in some with great perseverance through the ages. Our nucleated-cell ancestral prodigies (which could swim, metabolize sugar, regulate salts, breathe oxygen, produce H_2S gas, and take in live bacterial food) evolved because of their exquisite sensitivity: attraction to sugars and each other, struggle, fusion, eventual incorporation, and integration by compromise. Our sensibilities come directly from the world of bacteria.

We malign our ancestors with our harsh words. Our intolerant slogans denigrate the nonhuman life with which we share the planet. The bacte-

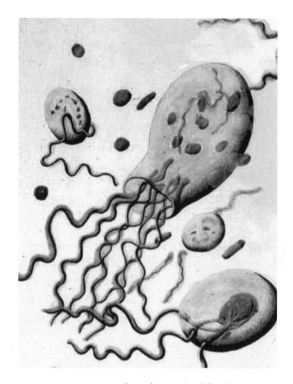

Figure 5.2 Spirochetes become undulipodia.
Drawing by Christie Lyons.

rial patina more likely will rid this planet of us, the voluble, ignorant ape, far sooner than we will cause any type of bacteria to disappear. No matter how we protest and what we proclaim, they most likely will thrive, frolicking to grow and reproduce in their own way, long after we perish.

Chapter 5 Notes

1. For detailed comprehensive explanation of the crucial, immense subvisible world of bacteria, see Madigan et al., 2003.
2. Margulis et al, 2006. Our video, *Eukaryosis* (L. Margulis and James MacAllister, 2005), shows there are no "missing links." All transitions that we hypothesize to be relevant to the origin of eukaryotic cells, i.e., to the evolution of protoctists from bacterial communities, are observable in extant organisms.
3. Case, 2008 explains kingdom vs domain and other recent issues of microbial taxonomy. Also see Margulis, 1998.

All for One

LYNN MARGULIS AND DORION SAGAN

Here, after you meet *Mixotricha paradoxa*, you will understand the chimeric nature of being and that all cells evolved from accumulation and genetic integration of once free-living symbionts.

In the heat of tropical Australia, near the northern town of Darwin in the national park called Kakadu, lives "mixed-up hairs paradoxical." Called *Mixotricha paradoxa*, this sedate swimmer is a microscopic organism comprising hundreds of thousands of smaller life-forms. *M. paradoxa* is

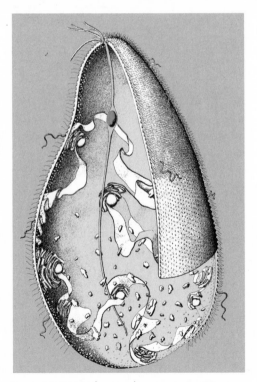

Figure 6.1 *Mixotricha paradoxa*. Drawing by Christie Lyons.

an extreme example of how all plants and animals—including ourselves—have evolved to incorporate masses. It swims around and around in the murky hindgut of its home: the rear part of the intestine of an ancient wood-eating termite called *Mastotermes darwiniensis*. Both insect and microbe *Mixotricha* are found nowhere else in the world!

The hullabaloo over mapping the human genome—the sum of all the genes in an individual—might lead one to think that each species has only a single genome and that the genetic makeup of individual organisms is discrete and unitary. Such is far from the case. Paraphrasing Walt Whitman, we animals contain multitudes. All cells of animals have at least two interacting genomes. One is the DNA in the cell nucleus; this is the genome that has recently been "mapped." The other is that of the DNA in the mitochondria—the cell's multiple oxygen-breathing organelles that are inherited only through the mother's egg. For more than a century, some scientists have known that every animal is in fact a multiple being, but until recently these unorthodox researchers were ignored.

The pioneering Russian naturalist Konstantin Sergeeivich Merezhkovsky argued in 1909 that the little green dots in plant cells that make sugar in the presence of sunlight (chloroplasts) were originally separate organisms.[1] The animals we think we know best (mammals, reptiles, insects) have genomes that determine limbs, eyes, and nervous systems that, for example, are very similar to our own. Like us, they are doubly genomic. Even some unicellular beings, protists such as paramecia and amoebae that lack eyes, limbs, and nervous systems, contain both nuclear and mitochondrial genomes. (The protists are unicellular members of the Kingdom Protoctista. If you haven't heard of them, don't feel bad: most of their members are either unsavory or too small to be seen. The protoctists, however, which include slime molds, kelp, amoebae, and seaweeds, are important because it is in this motley group that sexual reproduction evolved. Indeed, all plants, animals, and fungi arose from the protoctists, which include single-celled protists and their multicellular descendants. The term "protoctist" was coined by John Hogg just before he died in 1861. In chapters 9 and 13 we will return to look more closely at these fascinating, still underappreciated, and little-known ancestors.)

Plants and algae have these double genomes as well, plus a third genome of symbiotic origin. During their evolutionary history, their ancestors ingested, but did not digest, photosynthetic blue-green bacteria. Therefore, all visible photosynthetic organisms have at least three genomes. But many live beings—such as the mixed-up hairy paradox,

Mixotricha from Australia, that inhabits the guts of *Mastotermes* termites—contain within them up to five or more genomes.

The great nineteenth-century naturalist Joseph Leidy, one of the founders
of the Academy of Natural Sciences in Philadelphia, was the first in North
America to take a close-up look at the contents of a termite's intestine. "In
watching the Termites from time to time wandering along their passages
beneath stones," he wrote in 1851, "I have often wondered as to what
might be the exact nature of their food." What he saw under his microscope
amazed him. If the termite's intestine is ruptured by the experimenter, he
wrote, "myriads of the living occupants escape, reminding one of the
turning out of a multitude of persons from the door of a crowded meeting-
house." Leidy immediately realized that what he knew as "white ants" were
actually composed of dozens of different kinds of tiny life-forms, including
bacteria and what we now know are protists.[2] We now recognize that the
immense and motley crew that Leidy observed within a termite is in no way
a gratuitous add-on or a pathological infection.[3] Rather, it is a necessary
part of the termite's digestive system and is organized as a particular tissue.
The aggregate turns the refractory compounds lignin and cellulose, the
main constituents of wood, into food. This composite fabric, or living con-
sortium, has evolved in the nearly oxygen-free closed system of the termite's
abdomen for over one hundred million years. In only three weeks without
the living, wood-degrading, busy communities that have become their
digestive systems, the termites starve.

Merezhkovsky, when he argued that the chloroplasts of plant cells
evolved from symbionts of foreign origin, proposed the term *symbiogenesis*
for the merger of different kinds of life-forms into new species. Boris Kozo-
Polyansky, his successor, showed clearly in 1926 that symbiogenesis is a
major creative force in the production of new kinds of organisms.[4] Another
Russian botanist, Andrey S. Famintsyn, a plant physiologist, and an
American anatomist, Ivan E. Wallin, worked independently during the early
decades of the twentieth century on similar hypotheses.[1] Wallin further
developed his unconventional view that all kinds of symbioses played a crucial role in evolution. Famintsyn, believing that chloroplasts were symbionts,
succeeded in maintaining them outside the plant cell. Both men experimented with the physiology of mitochondria, chloroplasts, and bacteria and
found striking similarities in their structure and function. Chloroplasts, they
proposed, originally entered cells as live food—microbes that fought to survive—and were then exploited by their ingestors. They remained within the
larger cells down through the ages, protected and always ready to reproduce.

Famintsyn died in 1918; Wallin and Merezhkovsky's research work was rejected by their fellow biologists.[1] Their studies were ridiculed and nearly forgotten. Recent molecular biology has proved, however, that the cell's two sets of membrane-bounded organelles—chloroplasts in plants and mitochondria in both plants and animals—are highly integrated and well-organized former bacteria. The question of how these bacteria became permanent symbionts has been resolved, at least in outline.

Acceptance of the composite nature of the individual revolutionizes our concepts of evolution. Bacteria are exemplary genetic engineers: splicers and dicers and mergers of genomes par excellence. We people just borrow their native skills.

Like nearly all other animals, we mammals harbor in our intestines an assortment of specific bacteria that help us digest our food. Some can and others can't live outside humans. Without these hitchhikers to help digest fiber and produce vitamins, we—like termites—weaken and even die. Entirely integral to our bodies, however, are the mitochondria inside our nucleated cells. These tiny entities use oxygen to generate the chemical energy needed to sustain life. They reproduce on their own, independently of nuclear DNA, and multiply more quickly after short bursts of muscular exercise, leading to stronger, more heavily mitochondria-packed muscles. Because mitochondria are so genetically integrated into each of our cells, no one has yet succeeded in growing them in test tubes.

We believe that Wallin and Merezhkovsky were fundamentally correct when they claimed that all nucleated living things evolved by symbiogenesis, generally because of preexisting bacterial genomes physically associated with other organisms. Reef-building corals, for instance, are now known to have five different genomes of once independent organisms. And *Mixotricha paradoxa*, a compound beauty found in the gut of *Mastotermes*, also has five genomes. Indeed, *M. paradoxa* is the "poster protist" for symbiogenesis.

The Australian zoologist J. L. Sutherland first described and named "the paradoxical being with mixed-up hairs" in 1933. She reported it as a protozoan cell that swims by simultaneous use of both flagella and cilia. Since Sutherland's discovery, this organism has been subjected to many years of study and photography. Under low magnification, *M. paradoxa* looks like a single-celled swimming ciliate. The electron microscope, however, reveals that it consists of five distinct kinds of creatures. Externally it is most obviously the kind of "large" one-celled organism that is classified as a protist. Studies done in the 1950s by A.V. Grimstone of Cambridge University in the United Kingdom and the late L. R. Cleveland of Harvard showed that

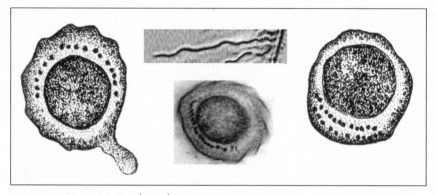

Figure 6.2 *Canaleparolina*. Photographs in the center; Drawings by Christie Lyons.

M. paradoxa is a hundred times larger than its close relative, *Trichomonas vaginalis*, so well known for causing vaginal itch. But inside each nucleated cell, where one would expect to find mitochondria, are instead many spherical bacteria. Where cilia should be, are some 250,000 hairlike *Treponema* spirochete bacteria; they greatly resemble the type that causes syphilis. A contingent of large rod bacteria is also 250,000 strong. In addition, we have redescribed some 200 larger spirochetes per *Mixotricha* cell. We named them *Canaleparolina*, after a beloved professor, Ercole Canale-Parola at the University of Massachusetts Amherst, who taught about spirochete bacteria for years.

Devoid of immune systems, well-fed bacteria can always reproduce without any mates. Yet they are supremely promiscuous beings in which infection and sex—that is, gene flow—are virtually the same thing. The sexual proclivities of bacteria include, when their survival is threatened, rampant donation, and reception, of genes.

Eventually we expect that the sum-of-themselves "neoDarwinist" evolutionists will realize with us, and Wallin and our Russian predecessors,[4] that natural selection operates not so much by acting on random mutations, which are often harmful, but on new kinds of individuals that evolve by symbiogenesis. Scrutinizing life at the microscopic level is like moving ever closer to a pointillist painting by Georges Seurat: the seemingly solid figures of humans, dogs, and trees, on close inspection, turn out to be made up of innumerable tiny dots and dashes, each with its own living attributes of color, density, and form.

Chapter 6 Notes

1. A summary of the early symbiogeneticists is in Margulis, 1990; details of the science and its history are described by Khakhina, 1992.

2. For Joseph Leidy's beautifully illustrated, accurate nineteenth-century work see Margulis, 2005.
3. Leidy's spore-forming intestinal bacteria, his "jointed threads" are generally harmless. When "weaponized" however, by acquisition of certain genes, for example, for toxin production, they become the "anthrax bacterium": *Bacilllus anthracis*. See our work based in large measure on Jeremy Jorgensen's master's thesis in Margulis et al., 1998.
4. Kozo-Polyansky, 1924 (in Russia).

Speculation on Speculation

LYNN MARGULIS

Here I speculate on the spirochete origin of our sensory-nervous systems. A strange idea, it is easy to resist because it seems so bizarre. But because life's biochemistry and genetics are so conservative it is, I suspect, correct.

Whereas in science theory is lauded, speculation is ridiculed. A biologist accused in print of "speculation" is branded for the remainder of her career. This biologist finds herself like a ballet dancer imitating a pigeon-toed hunchback: all of the intellectual training to keep my toes turned out emotionally backfires with any request to speculate freely.

In a manuscript deficient in references and lacking data, field and laboratory observations, and descriptions of equipment and their correlated methodologies, I feel a huge restraint as I attempt to slacken the bonds of professionalism and turn in my toes. My inhibitions fade with this opportunity to tell you what I really think!

We all intuit the reality of these terms: *perception*, *awareness*, *speculation*, *thought*, *memory*, *knowledge*, and *consciousness*. Most of us would claim that these qualities of mind have been listed more or less in evolutionary order. It is obvious that bacteria perceive sugars and algae perceive light. Dogs are aware; when deciding whether to chase a ball or not, they seem to be speculating. Thought and memory are clearly present in nonhuman animals such as *Aplysia*, the huge, shell-less marine snail that can be taught association. *Aplysia*, known as the sea hare, has been trained to anticipate; it will flee from potential electric shock as soon as a light is flashed. Knowledge, some admit, can be displayed by whales, bears, bats, and other vertebrates, including birds. But conventional wisdom tells us that consciousness is limited to people and our immediate ancestors. Many scientists believe that "mind"—whatever it is—will never be known by any combination of neurophysiology, neuroanatomy, genetics, neuropharmacology, or any other materialistic science. Brain may be knowable by the "ologies," but mind can never be.

I disagree with many versions of this common myth. I think brain is mind and mind is brain, and that science, broadly conceived, is the most effective method for learning about mind-body-brain. The results of the "–ologies" just listed, as well as of many other sciences, can tell us about ourselves and what is inside our heads. Furthermore, humans have no monopoly whatsoever on these mental processes. As long as we indicate consciousness of what, I can point to conscious, actively communicating, pond-water microscopic life (and even extremely unconscious bureaucrats). The processes of perception, awareness, speculation, and the like evolved in the microcosm: the subvisible world of our subvisible ancestors. Movement itself is an ancestral bacterial trait, and thought, I am suggesting, is a kind of cell movement.

We admit that computers have precedents: electricity, electronic circuits, silica semiconductors, screws, nuts, and bolts. The miracle of the computer is the way in which its parts are assembled. So, too, human minds have precedents; the uniqueness is in the recombination and interaction of the elements that comprise the mind-brain. My contention is that hundreds of biologists, psychologists, philosophers, and others making inquiries of mind-brain have failed to identify even the analogues of electricity, electronic circuits, silica semiconductors, screws, nuts, and bolts. In the absence of knowing the parts and their meanings, we can never know the human mind-brain. Only the very recent history of the human brain is illuminated by comparative studies of amphibian and reptilian brains. The crucial ancient beginnings of the human brain lie in the dancing of bacteria: the intricate mechanisms of cell motility. How do cells locomote? The answer to this puzzle is the beginning of enlightenment for the origins of mind-brain.

I cherish a testable, scientific theory. Many means for testing it include biochemical, genetic, and molecular-biological techniques. The facilities for verification of the idea are available in New York City. A conclusive proof would have required generosity on the part of at least three highly talented scientists, and their laboratory assistants.[1] Charles Cantor, formerly of Columbia University Medical School, and David Luck and John Hall, geneticists at Rockefeller University, have developed techniques to purify genes (DNA) gently. The purification holds the biological material on blocks of agar (a gelatin-like substance) in such a manner that the structures in which the genes reside, the chromosomes, are extracted in their natural long, skinny form. Groups of genes, linkage groups, can be identified. Chromosome counts, difficult to determine microscopically, can be made biochemically.

Professor David Luck, and his colleague Dr. John Hall, both geneticists at Rockefeller University for over a quarter of a century, discovered in the green algae *Chlamydomonas* a new type of genetic system.[2] They found a set of genes that determine the development of kinetosomes, intracellular structures present in very different kinds of motile cells, those of this and other green algae, sperm (see figure 5.1, p. 40), ciliates, oviducts, and tracheae, for example. Kinetosomes, which I think of as assembly systems for cell motors, are apparently determined by a unique set of genes separable from those of the nuclei and other cell components. These genes, as inferred from genetic studies of Luck and Hall, may be the spirochetal remnant genes I predicted must be inside all cells that contain kinetosomes.

Although no exorbitant amount of money would be needed, the tests of my theory are limited. They require time and energy of very busy people. Furthermore, the results of tests, even if the scientific research were ideal, would cure no disease, stop no war, limit no radioactivity, save no tropical forest, and produce no marketable product—at least at first. There would be no immediate profit from the work. The results would simply help reconstruct the origin of our mind-body's sensory-nervous system from its bacterial ancestors.

What is the idea? I hypothesize that all the phenomena of mind, from perception to consciousness, originated from an unholy microscopic

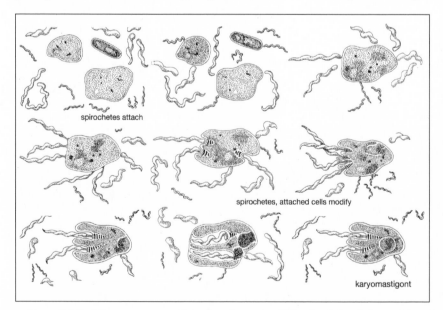

Figure 7.1 Evolutionary sequence: Spirochetes become undulipodia. Stripes in lower right represent kinetosomes at the base of the mobility structures. Drawing by Kathryn Delisle.

alliance between hungry swimming killer bacteria and their potential archaebacterial victims.[3] The hungry killers were extraordinarily fast-swimming, skinny bacteria called spirochetes (figure 7.2). These active bacteria are relatives of the spirochetes of today that are associated with the venereal disease that, in prolonged and serious cases, infects the brain: the treponemes of syphilis. The fatter, slow-moving archaebacteria were quite different from the spirochetes. In resisting death the archaebacteria incorporated their fast-moving would-be killers into their bodies. The archaebacteria survived, continuing to be infected by the spirochetes. The odd couple survived. The archaebacteria were changed: they were made more motile, but not killed, by their attackers.

The antics of spirochetes in nature, photographed live through a microscope. Whether from the hindguts of termites, the digestive system of clams, the Muddy River at the fens in Boston, Massachusetts, or the salt flats near the delta of the Ebro River in northeast Spain (between Valencia and Barcelona), these microbes carry on their sensuous and social lives.

Our cells, including our nerve cells, I maintain, are latter-day products of such mergers—the thin, translucent bodies of the spirochete enemies sneakily incorporated inextricably and forever. The wily and fast movement, the hunger, and the sensory ability of the survivor's enemies were all put to good use by the evolving partnership. Cultural analogues of

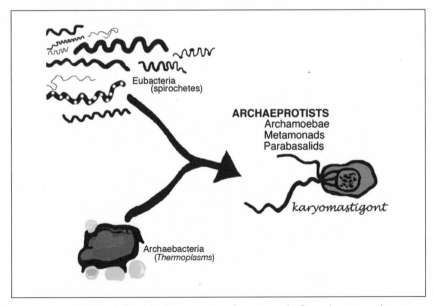

Figure 7.2 Bacteria form the *karyomastigont* by merger; the first eukaryote evolves.

such mergers exist, as cases in which two very different warring peoples form new identities after the truce. In such new identities, for example, the unique domesticated plants of one culture may become firmly incorporated into that of the second, just as the presence of Indian corn, tomatoes, and potatoes in Europe is due to the near annihilation of indigenous Native Americans. I see our cell movement, including the movements leading to thought, as the spoils of ancient microbial battles.[4]

My speculations, two thousand million years later, may be the creative outcome of an ancient uneasy peace. If this reckoning is true, then the spirochetal remnants may be struggling to exist in our brains, attempting to swim, grow, feed, connect with their fellows, and reproduce. The interactions between these subvisible actors, now full member-components of our nerve cells, are sensitive to the experience we bring them. Perception, thought, speculation, and memory, of course, are all active processes; I speculate that these are the large-scale manifestations of the small-scale community ecology, that is, the fusion of two ancient forms of bacteria.

Arcana Naturae Detecta is the name of Anton van Leeuwenhoek's seventeenth-century collected letters revealing the microcosm beneath his single-lens microscope illuminated by a gas lamp. The visible became explicable to him by the machinations of the subvisible. Leeuwenhoek and his followers made clear that "decay," "spoiling," and "rotting food" are all signs of healthy bacterial and fungal growth. In baking, "rising dough" is respiring yeast; in tropical disease, malarial fevers are apicomplexan protists bursting our red blood cells. Fertility is owed in part to semen or "male seed" containing millions of tailed sperm in the sugary solution of semen. The disease of Mimi, the heroine of *La Bohème*, is "consumption." From its point of view, "consumption" is the healthy growth of *Mycobacterium* in the warm, moist lungs of the lovely young woman. Speculation, I claim, is the legacy of the itching enmities of unsteady truce. Speculation is the mutual stimulation of the restrained microbial inhabitants that, entirely inside their former archaebacterial enemies, have strongly interacted with them for hundreds of millions of years. Our nerve cells are the outcome of an ancient, nearly immortal marriage of two archenemies who have managed to coexist: the former spirochetes and former archaebacteria that now comprise our brains.

Like animated vermicelli married and in perpetual copulatory stance with their would-be archaebacterial victims, these former free-living bacteria became inextricably united by the process illustrated in figure 7.2. They may have been united for well over two-thousand million years. The fastidiously described speculation is indistinguishable from the theory. I continually play with an idea: the origin of thought and consciousness is cellular, owing its beginnings to the first courtship between unlikely bacterial bedfellows who became ancestors to our mind-brains.

My goal in the rest of this essay is to explain what I mean and why I make such an extraordinary assertion.

What needs to be explained? My basic speculation is that mind-brain processes are the nutrition, physiology, sexuality, reproduction, and community ecology of the microbes that compose us. The microbes are not just metaphors; their remnants inhabit our brain, and their needs and habits, histories, and health status help determine our behavior. If we feel possessed and of several minds, if we feel overwhelmed by complexity, it is because we are inhabited by and comprised of complexities.

The detailed consequences of the theory of spirochete origin of microtubules of brain cells do not belong in a popular essay about speculation. Indeed, our limited statement has been published in the august professional journal *The Proceedings of the National Academy of Sciences.*[5] Yet, I ask that the formerly unmentionable become widely discussable by specialist scientists and inquisitive philosophers so that the consequences of the hypothesis may be explored. Could thought, speculation, and awareness really have evolved from fast-moving bacteria and their interactions, their hungers, their activities, their satiations, their associations with their fellows, both like and unlike, and their waste-removal processes? Is it possible that we are as entirely unaware of the microbial inhabitants that comprise us as a huge ship tossing in the waves is unaware that her responses are determined by the hunger, thirst, and eyesight of the captain at the helm and his communications with the crew.

Two kinds of spirochetes at least, large *Canaleparolina* and small treponemes (like those associated with syphilis) and others, are permanently attached to *Mixotricha paradoxa*, the microbe found in a wood-eating termite from Australia (see chapter 6). Simultaneous movement by hundreds of attached spirochetes make *M. paradoxa* swim forward.

What might be the implications for mind-brains if this bacterial origin of speculation is correct? Let us list a few. They all may be incorrect, but they are all testable within the rigors of the scientific tradition.

1. Nerve impulses and the firing of nerves. These become explicable as our motile spirochetes' struggle to swim; as Betsey Dyer (biology professor at Wheaton College, Massachusetts) says, captive former spirochetes are spinning their wheels, unable to move forward. They have

become uncoupled motors going around and around.[6] This quasi-movement is the nerve impulse. It occurs because small, positively charged ions (for example, sodium, potassium, calcium) are accumulated and released across what is now our nerve-cell membrane. These ions, their protein and membrane interactions, derive from fusion of membranes from both the original archaebacteria and the original spirochetes.

2. Sweet memories. Two different kinds of memory systems exist: short term (seconds to minutes) and long term (indefinite). The storage of memories is markedly enhanced by adrenaline and other substances that lead directly to increased availability of sugar to the brain cells. Sugar, like any substance that penetrates the brain—that is, that enters the brain from the blood—is very carefully monitored and controlled.[7]

Short-term memory arises every time from casual encounters between the sticking-out parts of former spirochetes and their friends. These interactions begin in seconds; it probably takes a few minutes at most while two or more neurons, née spirochetes, interact. The casual encounters occur by small-ion interactions with proteins on the surfaces of what used to be spirochete membranes (now they are our nerve-cell membranes). In brief, short-term memories derive from the physiology of spirochetal remnants in the brain. We know that the pictorial short-term memory, for the recognition of fractal designs, for example, "is coded by temporary activation of an ensemble of neurons in the region of the association cortex that processes visual information." The "temporary activation," if I am correct, will be directly homologous to spirochete behavioral interaction—not analogous to it or to computer-software manipulation.

Long-term memory is stable; it depends on new protein synthesis. Long-term memory works because it stores the short term. What were repeated casual encounters between former spirochetal remnants become stabilized attachment sites. "Synapse," if I am correct, is the neurophysiologist's term for the well-developed spirochetal remnant site of interaction. In brief, long-term memories derive from the growth of spirochetal remnants, including their attachment sites, in the brain.

Sugar enhances memory processes because it feeds preferentially the spirochetal remnants so that they can interact healthfully and form new attachments. Sugar has been the food of spirochetes since they squiggled in the mud.

As Edelman[8] has pointed out, no two monkeys, no two identical twins, are identical at the level of fine structure of their neuronal connections. "There must be a generator of diversity during the development of neural circuits, capable of constructing definite patterns of groups but also generating great individual variation. Variation must occur at the level of cell-to-cell recognition by a molecular process. Second, there must be evidence from group selection and competition in brain maps and re-entrant circuits. This must occur not in the circuitry but in the efficacy of preformed connections or synapses." I believe Edelman is discovering the actively growing latter-day populations of former microbes that comprise every brain. Edelman's "populations" are nerve cells and their connections. I interpret Edelman's populations literally as remnants of ancestral microbial masses. The spirochetal remnants, either poised or ready to grow, attach and interact depending on how they are treated during a human's crucial stages of fetal development, infancy, and early childhood. Neural Darwinism, differential growth by selection of spirochete associations, determines the way in which the brain develops.

Mental health is, in part, how we feed the healthy spirochetal remnants that make up our brain. Learning becomes a function of the number and quality of new connections—interactions and attachments—that these wily former bacteria forge. The spirochetal remnants grow more quickly, dissolving temporary points of contact while consolidating firm connections that are our nerve-cell endings, during our infancy and childhood. More potential changes occur early—in infancy and adolescence—relative to those of adulthood. The growth patterns of nerve cells née spirochetes are sensitive to the food, such as essential fatty acids, that the rest of our body provides for them; experience is always active, always participatory, and, if registered in long-term memory, unforgotten. Our memories are the spirochetal remnants' physical networks. Our crises and climaxes are their "blooms," their population explosions. Senility is spirochetal-remnant atrophy. It is no coincidence that salt ions and psychoactive drugs, including anesthetics, have strong effects on movement or growth of the free-living mud-bound cousin spirochetes.

Clearly these enormous contemplative issues cannot be solved here by me. All I suggest is that we compare consciousness with spirochete microbial ecology. We may be vessels, large ships, unwitting sanctuaries to the thriving communities comprising us. When they are starved, cramped, or stimulated we have inchoate feelings. Perhaps we should get to know ourselves better. We might then recognize our speculations as the dance networks of ancient, restless, tiny beings that connect our parts.

Chapter 7 Notes

1. I say "would have" because David Luck (1932–1999) died before this work was planned and undertaken. As he was director of research at the extremely competent and powerful Rockefeller University, if I had communicated with him we might have undertaken the relevant research properly. I have not given up. With some aid from enlightened private donors, John Hall and I continue to seek the spirochete contribution to eukaryotic cells. New results are encouraging. (Hall and Margulis, 2008.)

2. Hall et al., 1989.

3. Please see the *Readings* that begin on page 238 for these references: Margulis, L. 1993; Margulis, 1999; Margulis and Sagan, 2002; and Margulis et al., 2006. They detail these ideas from the popular to the professional level.

4. Margulis and Sagan, 1997.

5. Margulis, et al. 2006.

6. For Betsey Dexter Dyer's published work please see the *Readings* that begin on page 238.

7. Miyashita and Chang, 1988.

8. Edelman, 1985.

Spirochetes Awake: Syphilis and Nietzsche's Mad Genius

LYNN MARGULIS

I have not been able to locate acceptable scientific evidence in the published professional literature that the human immuno-logical retrovirus (HIV) causes AIDS. Rather, with Duesberg, I conclude the claim that "HIV causes AIDS" is an invention.

The parallels of acquired immunological syndrome symp-toms with those presented by syphilis are astonishing. Perhaps there are no new diseases, only new drugs.

In the foothills of the Italian Alps, on a snow-draped piazza in Turin, on January 3, 1889, a driver was flogging his horse when a man flung his arms around the poor beast's neck, his tears soaking its mane. The horse's savior was the German philosopher Friedrich Wilhelm Nietzsche (1844–1900). His landlord later found him collapsed in the square and brought him back to his room, where Nietzsche spent the night writing a flurry of bizarre postcards. As soon as his friend and colleague Jacob Burckhardt received one of these crazed letters, he convinced his close friend Peter Gast to go and accompany Nietzsche on his return to Basel. Much of the rest of the century, the last eleven years of his life, Nietzsche spent in incoherent madness, crouching in corners and drinking his urine. The most productive year of his career had been immediately prior to the psychotic break. After it, he wrote no more philosophy. Deborah Hayden summed up the famous incident:

> The story of Nietzsche's sudden plummet from the most advanced thought of his time to raving dementia is often told as if there were a razor's edge demarcation between sanity and tertiary syphilis, as if on 3 January armies of spirochetes woke suddenly from decades of slumber and

Figure 8.1 Friedrich Nietzsche. Courtesy of Marti Dominguez and Christie Lyons.

attacked the brain, instead of the biological reality that
paresis is a gradual process presaged over many years.[1]

Hayden's case to prove that Nietzsche indeed suffered all his adult life
from syphilis is as strong as any posthumous medical history can be. He
was diagnosed at a time when clinical familiarity with the disease
abounded. Detailed evidence shows that he passed through each of the
three stages: the chancre of primary syphilis immediately after infection;
the terrible pox, fever, and pain of secondary syphilis that emerges from
months to years later; and the dreaded third, "paresis." *PARESIS*, like the
word *syphilis* itself, refers to a syndrome. An acronym, it expands to
*Personality disturbances, Affect abnormalities, Reflex hyperactivity, Eye
abnormalities, Sensorium changes, Intellectual impairment,* and *Slurred*

Figure 8.2 Pox inspection of the ladies of the night. Courtesy of Marti Dominguez.

speech. Paresis often begins with a dramatic delusional episode, but in the following months and years, dementia alternates with periods of such clarity that there seems to have been a cure.

Infection by the spirochete of syphilis—declared eradicated by the mid-twentieth century—still prevails, I believe. The efficacy of penicillin for early treatment, improved hygiene, condom use, and attitudes that lead the afflicted to seek help for venereal infection conspire to bolster the common myth that syphilis has disappeared. We are deceived; I think that many people suffer from syphilis called by other names.

Syphilis symptoms are caused by venereal infection with a spirochete bacterium called *Treponema pallidum*. The treponeme family of spirochetes consists of corkscrew-shaped bacteria, all of which swim and grow in animal tissue. The bacterial flagella, encased within an outer membrane, are inside the cell; for that reason they are called "periplasmic flagella." Spirochetes, like other "gram negative" bacteria, all have two cell membranes with a space between them. In this "periplasmic" space between the inner and outer membranes the flagella rotate.[2] Smaller spirochetes such as the syphilis treponeme have only two to four such flagella, whereas some giant spirochetes have more three hundred. The efficient screw-wise motion into genital and other tissue requires this flagella arrangement. The confusion and misinformation about syphilis symptoms, contagion, viruses and "germs" like spirochetes have abounded for hundreds of years.[3]

Treponema pallidum is one freak among a huge diversity. The vast majority of spirochetes live peacefully in mud, swamps, and waterlogged soils all over the world. Benign, "free-living" spirochete relatives of *Treponema pallidum* are everywhere. They thrive where food is plentiful: lake shores rich in decaying vegetation, marine animal carcasses, hot sulfurous springs, intestines of wood-eating termites and cockroaches, and the human mouth. Most kinds are poisoned by oxygen, which they swim away from to avoid. Very few cause illness. Nevertheless, ticks infected with the *Borrelia burgdorferi* spirochete of Lyme disease can induce serious arthritis and other enduring symptoms. In Europe Lyme disease is called "erythema migrans." In Australia it is known as "tick arthritis." Yet another spirochete nearly indistinguishable from the Lyme disease *Borrelia* is a healthy symbiont in the intestines of termites. A treponeme similar to that of syphilis is associated with yaws, an eye disease of tropical climates. Leptospirosis, a systemic and sometimes fatal infection found usually in fishermen, is due to spirochetes that are carried in the kidney tubules of rats that urinate into nearby water. The fishermen acquire *Leptospira* spirochetes from fish hook cuts and other skin lesions. And, of course, there is syphilis.

Nietzsche's letters from 1867 until his breakdown provide a vivid account of the suffering of secondary syphilis. He complains of the pain, skin sores, weakness, and loss of vision that typify the repertoire of the disease. In his last year, his letters give evidence of euphoria. His published works show the grandeur and inspiration that tertiary syphilis sometimes brings to brilliant and disciplined creative minds by removing inhibition as brain tissue is destroyed. In *Thus Spoke Zarathustra* (1884) Nietzsche wrote, "Die Erde, sagte Er, hat eine Haut; und diese Haut hat Krankheiten. Eine diese Krankheit heist zum Beispiel: 'Mensch.'" This has been translated as "The Earth, he says, has a skin, and this skin has a sickness. One of these sicknesses is called 'man.'" Or "The Earth is a beautiful place but it has a pox called man." What terrible insight Nietzsche must have had into the devastating horror of pox!

Multiple sources indicate that he was treated for syphilis in 1867 at the age of twenty-three. Seeking medical treatment for eye inflammation, a frequent syphilitic symptom, he consulted Dr. Otto Eiser, who reported not only Nietzsche's penile lesions but that he had engaged in sexual relations several times on doctor's orders! Years later, in 1889, when Nietzsche broke down and was taken to the clinic of a paresis expert, Hayden (2003) tells us that he was admitted with the diagnosis "1866. *Syphilit. Infect.*"

In 1888 Nietzsche's productivity was, by any standard, extraordinary. He completed his philosophical project *Twilight of the Idols*, *The Antichrist*, *Ecce Homo*, and *The Case of Wagner*. The style of these works is apocalyptic, prophetic, incendiary, and megalomaniacal, leading many scholars to claim the excesses of these works was due to incipient paresis.

Now, after more than half a millennium of the study of syphilis and more than a century after Nietzsche's breakdown, our research suggests that the philosopher really did plummet abruptly into madness; armies of spirochetes *did* awaken suddenly from decades of slumber and literally began to eat his brain.

Many claim syphilis was known in Europe prior to the return of Columbus, but as Hayden describes and I agree, it is more likely the insidious venereal infection was a new gift of the Americas to the people of Europe. Columbus and his crew returned to Spain with a novel set of symptoms that soon spread to Naples and France. From that first year, 1493, the disease was described in detail, beginning with the physician who treated Columbus and his men, Dr. Ruiz Diaz de Isla. Diaz de Isla reported, "And since the Admiral Don Cristobal Colon had relations and congress with the inhabitants . . . and since it is contagious, it spread." Eventually it affected the waterfront prostitutes of Barcelona. Diaz, in work published in 1539, wrote that infected sailors were accepted both into the army that Charles of France brought to besiege Naples in 1495 and into the forces Ferdinand of Spain employed to defend Naples. Ferdinand's army alone is estimated to have had five hundred prostitutes among its camp followers. Soon after the victorious entry of Charles's army, the Great Pox of Naples erupted. Charles himself returned to France infected. His multinational mercenaries brought infection back to every European country. By the next year, the disease spread across the continent, puzzling physicians with its novelty.

Within the first few decades of the contagion, in cities across Europe physicians reported that between 5 and 20 percent of the population suffered. Variously named at first, it came to be called *morbus gallicus*, the French malady. Charles's army was blamed for its introduction to Naples—perhaps rightly. Physicians, who published in the lingua franca of Latin, soon after the disease's great outbreak in 1495, drew international attention. Girolamo Fracastoro, in 1530, wrote a verse treatise on the disease entitled *Syphilus sive Morbus Gallicus* in which the eponymous protagonist, a shepherd, is the first to bear the disease, as a punishment for impiety. (*Natural History*, October 2000, published an article on Fracastoro, one of the last by Stephen J. Gould.) The name stuck.

Syphilis has been surprisingly well documented since its outbreak in the closing years of the fifteenth century, as microbiologist and sociologist of science Ludwik Fleck (1896–1961) wrote in his masterpiece about the genesis and development of scientific facts.[3] At least eight treatises of syphilology have survived from the years 1495 to 1498 alone—the first three years after its explosive spread. The earliest conference led by Nicolò Leoniceno, professor of medicine at the University of Ferrara, the physician who performed autopsies on victims of the Pox of Naples, was held in 1497 in Toulon, France. From the sixteenth century through the end of the nineteenth century the prevalence and peculiarities of syphilis led to publications from scientific arcana to torrid novels.

The cause of the disease was avidly sought. In 1905 Erich Hoffmann sent a genital chancre specimen to German microscopist Fritz Schaudinn, who confirmed the etiology. He aptly called the lively, translucent, thin, corkscrew-shaped bacterium he observed "thin, pale thread": *Treponema pallidum*. *Treponema pallidum* spirochetes were found in the brains of patients that manifested tertiary syphilis symptoms by Udo J. Wile in 1913.

Syphilis has gained attention again because of its disputed relationship to AIDS. Today, although physicians rarely record cases of tertiary syphilis, the earlier two stages of the disease seem on the rise. AIDS patients who had a record in their past of syphilis and who were apparently cured by antibiotics succumb again to syphilis. "Syphilis in patients infected with HIV is often more malignant with a greater disposition for neurological relapses following treatment," the scientist, not a physician, Dr. Russell Johnson of the University of Minnesota medical school, a world expert on *Borrelia burgdorferi*, the Lyme disease spirochete, said to me in his usual cautious manner. Johnson is a laboratory microbiologist who capably grows *Borrelia* spirochetes outside any animal bodies and free of other types of microbial life.

Dr. Peter Duesberg, one of the discoverers of the machinations of the retroviruses, rejects exclusive focus on HIV as the cause of AIDS.[4] In his excellent book *Inventing the AIDS Virus*, he questions a prevalent assumption that, as a contagious virus, HIV is even the main cause of the lesions, tumors, rashes, arthritis, weakness, pneumonia, and other severities that accompany immunosuppression. He only requests that we reevaluate our entrenched ideas and preconceived notions. These symptoms, including the detection in tissue of both the HIV antibody and the virus itself, may, as in other opportunistic infections, be the consequence he suggests, not the sole "cause of AIDS." The symptoms of immunosupression may betray the tenacity of the syphilis treponeme and correlate with the sexual and other behaviors of the patient.

Joan McKenna, a physiologist with a thermodynamic orientation from Berkeley, California, who has studied venereal disease for many years, wrote in a personal letter to me after a long telephone conversation:

> Because spirochetes can be harbored in any tissue for decades and can move from latency to reproductive stages, their survival in any host and despite any known therapy is nearly certain. . . . [We also] know that unknown factors will activate the microorganism [*Treponema pallidum*] from latency into an aggressive infection. . . .

She went on to remark, in regard to the intriguing (to say the least) relationship between syphilis and AIDS, that "no symptoms show up in AIDS that have not historically shown up with syphilis and the history of these populations [where AIDS is rampant] includes a high incidence of syphilis."

Clinical confusions, misdiagnoses, anomalous symptoms, conflated multiple infections, abound from 1495 in the early work of syphilology until now. Yet many studies confirm the variety and severity of symptoms attributable to the *Treponema pallidum* spirochete. The malady remains idiosyncratic in its course, with variability in the timing of the stages and, even now, the absence of any reliable test or single diagnostic. Still, the evidence suggests that the virulence and severity of the disease have diminished dramatically since the initial violent "pox" outbreak. No anomaly here needs explanation; rather, this behavior is expected of pathogens in first exposure to naive populations. Syphilis in Europe showed the same pattern that measles and smallpox did when first introduced to the Americas by Europeans. As early as the first few decades that followed the Pox of Naples, subsequent generations of Europeans were more resistant. Pathogenic microbes maximize not by rapid lethality but by conversion to chronic disease that lasts a lifetime and subtly affects behavior in the stricken animal.

Since the late nineteenth century the Wassermann blood test has often been touted as the best diagnostic test for syphilis. The fear of syphilis transmission was once so common that the Wassermann test was, and often still is, legally mandated in many places, required prior to marriage. However, as shown by Fleck[3] and others, the Wassermann reagent does not measure the presence of *Treponema pallidum*. It indicates, and not even 100 percent of the time, the exposure of a patient to unspecified infectious bacteria. The Wassermann test detects cardiolipin, a substance produced as a general healthy immune response. A Wassermann positive test shows that a person makes antibodies against certain blood-borne bacteria that may include the syphilis treponeme. Furthermore, the test in known syphilitics in advanced stages of the syndrome converts: it is negative. On another front, to preclude contagious mother-to-infant transmission of syphilis during parturition, drops of silver nitrate, thought to suppress the syphilitic spirochete, were placed in the eyes of most newborns. This practice occurs in some regions even now and even when blood tests for syphilis in the mother are negative. These irrational practices measure residual fear of the contagion of syphilis.

Arsphenamine, an arsenic-based remedy, was said to improve the health of syphilitic patients in the beginning of the twentieth century. Often it made people sicker. After 1943 came the "miracle drug": the claim was that a single or a few massive doses of penicillin cured the body permanently of the dreaded treponeme. After hefty antibiotic treatment in newly detected patients the insidious corkscrews disappeared. Though tiny, shiny round bodies, the apparent remains of "dead" spirochetes, might sometimes be found in tissue, the moving treponeme was declared gone. A researcher in Paris in the 1950s, J. Pillot, after whom the beautiful large spirochete *Pillotina* was named, "proved" that the round body remnants of the lively corkscrew are dead. The confusion comes from the fact that—penicillin or not—during the long latent phases of the disease after the primary chancre, moving corkscrew treponemes are not seen in tissue in any case. Many years and studies later we can say that whether or not any treponemes are visible in the patient, penicillin, except when given in appropriate dose very early in the course of the disease, is not an effective and permanent cure. The observation in 2006 that low doses of penicillin actually induces round bodies that emerge when the antibiotic is withdrawn has been detailed in the PhD dissertation by Andrei Belichenko, student of medical microbiologist Igor Bazikov from Stavropol, Russia.

Some physicians still insist that penicillin and strong immune systems definitively eliminate this disease, while others claim that treponemes "hide" in tissues inaccessible to antibiotics. Some speculate that tertiary syphilis occurs when the syphilis treponemes finally manage to spread, after decades of invisible stealth, and penetrate the "blood-brain barrier." Alas, most physicians and syphilis scholars (and scientists such as I) simply don't know the relationship between *Treponema pallidum*, syphilis symptoms, the immune response, secondary infection, sexual behavior, and the putative cures. A large body of Russian medical microbiological literature leads to the conclusion that no significant research has shown that syphilis can be cured with penicillin.

Finally, in 1998, the description of the entire genome of *Treponema pallidum*, at about 1,100 genes in total one of the smallest bacterial genomes known, was published. Two other spirochete genomes are known: that of the even smaller *Borrelia burgdorferi*, with some 900 genes, and that of *Leptospira*, with nearly 4,000 genes. Spirochetes like *Leptospira* that are capable of life outside the body of animals have at least five times as many genes as the *Borrelia* spirochete. The leptospires all by themselves internally produce all their necessary components (proteins, lipids, vitamins, and so on), whereas *Treponema pallidum* does very little by itself; it survives only on rich human tissue as its food. For this reason it is likely that both the *Borrelia burgdorferi* and the syphilis treponeme lost some four-fifths of their genes as they became "obligate parasites."

To identify any bacterium, the microbiologist needs to separate it and grow it by itself, that is, "in isolation." Despite the specific genome knowledge of the single treponeme strain investigated, the routine growth of any *Treponema pallidum* in isolation (outside the warm, nutritious mammalian body, usually rabbit testes) has not been achieved. Apparently many different treponemes exist in nature, including in human bodies, and a great array of different chemical components and environmental conditions are supplied by the surroundings. The *T. pallidum* does not cause illness in seriously infected rabbits. Whether in organic mud or changing human tissue, these spirochetes depend utterly on their immediate environment. No one who writes in English, as far as we know, has ever been able to induce round bodies of *Treponema pallidum* to form in isolation in a test tube and then, in isolation, to test them for viability, that is, for their ability to resume growth in tissue.

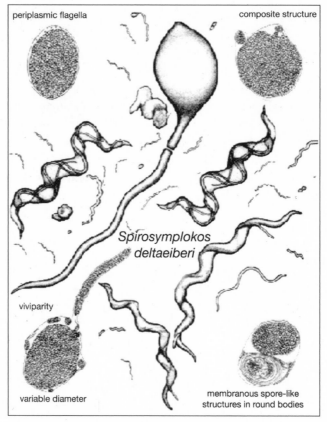

Figure 8.3 *Spirosymplokos deltaeiberi,* a microbial mat spirochete, life history.
Drawing by Christie Lyons; corner images are electron micrographs.

My students and colleagues and I are not experts on any disease bacteria, or even illnesses where symptoms are associated with visible spirochetes. Rather we have been living closely with spirochetes for very different reasons. Our interest is in the possible role these wily bacteria played in evolution of larger forms of life. Attempts to reconstruct the evolutionary history of the nucleated cell, the kind that divides by mitosis, has led us to study harmless spirochetes.

I suspect that the mitotic cell of animals, plants, and all other nucleated organisms (algae, water molds, ciliates, slime molds, fungi, and some fifty major groups included in the Protoctista kingdom) share a common spirochete ancestor. I believe that with much help from colleagues and students, we will soon be able to show that certain free-swimming spirochetes contributed their lithe, snaky, sneaky bodies to become both the ubiquitous mitotic apparatus and the familiar cilia of all cells that make such "moving hairs." Our lab work, coupled with that of other scientists, reveals that certain spirochetes when threatened by death can and do form immobile, shiny round bodies. Furthermore, these round bodies can hide and wait until conditions become favorable enough for growth to resume.

Since 1977 a group of scientists and students has been traveling to Laguna Figueroa (called Lake Mormona by anglophones) near San Quintin, Baja California Norte, Mexico, to study "microbial mats." These communities of organisms resemble ancient ones that left fossils in rocks. They are among the best evidence we have for Earth's oldest life-forms. Many times we brought microbial mat samples back to our lab and left them, as bottles of brightly colored mud, on the windowsill, where photosynthetic bacteria powered the community. On several occasions the bottles were assiduously ignored through semesters of classes and meetings. From time to time we took tiny samples from the bottles and placed them in test tubes under conditions favorable for growth. Spirochetes of different kinds did begin to swim and grow; we suspect they emerged from round bodies after they were put into fresh, clean, abundant liquid food. These spirochetes, mostly unidentified, persisted live but in hiding in these bottles and jars for at least ten years. We saw no active spirochetes in these jars for many months, indeed many years,

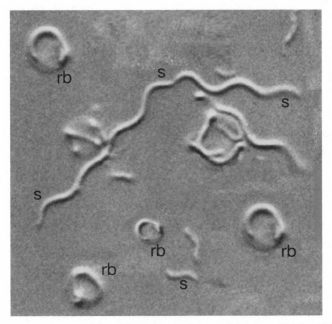

Figure 8.4 *Spirosymplokos deltaeiberi* spirochetes and their round bodies (rb).

after the samples were collected. However, when appropriate food was prepared and added to refampicin, an antibiotic that our mud spirochetes ignore but to which many other bacteria are sensitive, active swimming spirochetes were recovered.

Today we study another microbial community sample collected more than seventeen years ago, in 1990, by Tom Teal at Eel Pond, Woods Hole, Massachusetts. It is in our lab at the University of Massachusetts at Amherst in a 40-liter glass jar. We add no food and only "rain" (distilled water), but with sunlight as energy source an abundance of life still thrives. Long after no typical spirochetes were seen in any of these samples we added bits of either wet or dry mud to food and water known to support activities of spirochetes, swimming and growing. Within about a week "armies of spirochetes awoke" from at least months of slumber. As for more than one "decade of slumber" we still don't know.

We have observed and filmed spirochetes rounding up to form inactive bodies from all over the world—Cape Cod, Massachusetts, the Spanish Mediterranean coast, and the Pacific coast of Mexico. Continuation of work on spirochetes led to continued collaboration with Spanish colleagues, Mónica Solé and others with muds from the Ebro delta. Professors Ricardo Guerrero and Isabel Esteve had begun a strong research project. One stake, a stick in the mud labeled #1 UAB,[5] marks a site on a microbial mat that somehow seems exceptional. Many different fascinating organisms were taken from that place but none as interesting as the large spirochetes we named *Spirosymplokos deltaeiberi*. Whenever these easy-to-see spirochetes are confronted with harsh conditions such as liquid that does not support their growth, water that is too acid, sugars they cannot digest, or a temperature that is too high, they make round, dormant bodies like those Pillot and nearly all his successors argue are dead.

The spheres of *Spirosymplokos deltaeiberi* we studied live and by many other forms of microscopy look just like the round bodies published by Norwegian microbiologists Oystein and Sverre-Henning Brorson. (They call them cysts.) The Brorsons showed that under unfavorable conditions

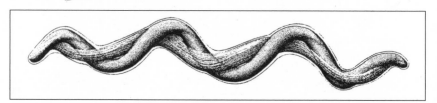

Figure 8.5 *Cristispira*, a large spirochete symbiotic in healthy oysters. Drawing by Christie Lyons.

the *Borrelia burgdorferi* spirochete of Lyme disease makes round bodies. After weeks of dormancy, no growth, and no sensitivity to antibiotics and other chemical insults, these round bodies revive. At high magnification they look just like those of *Spirosymplokos deltaeiberi*, only smaller. The *Borrelia burgdorferi* round bodies convert to form swimming spirochetes all at once and begin to grow easily as soon as they are placed into proper liquid food at the right temperature.

The Brorsons confirmed what we suspected: spirochete round bodies, like the spheres of *Spirosymplokos deltaeiberi*, are fully alive. Either mixed with other mud organisms or growing by themselves in isolation, just supply them with what they need to grow and within minutes they revert into swimming, active, feeding corkscrew spirochetes. Armies of them awake from months of slumber. Our work with Guerrero on *S. deltaeiberi* coupled with our reading of the literature, especially several studies by the Brorsons, leads us to emphasize an ancient secret of spirochete success: persistence via round bodies. The Russian investigators also find round bodies (cysts) and accept the concept of spirochete dormancy in tissue according to news from Victor Fet, Marshall University, Huntington, W.V.

Nietzsche's brain on January 3 acted like transfer of microbial-mat spirochetes into new, fresh food. Our interpretation is that they transformed from dormant round bodies to the swimming corkscrews in a very short time. Deborah Hayden, however, is also correct. Nietzsche was inoculated in his early twenties, and his longstanding condition was confirmed both by the physician's diagnostic on the medical record "*Syphilit. Infect*" and, at his death, by pox scars on his private parts. The wily spirochetes, many as dormant round bodies, had been living in his tissues for over thirty years. But on January 3rd in Turin hungry armies of revived spirochetes insinuated themselves into his brain tissue; the consequence was descent of Nietzsche the genius into Nietzsche the madman in less than one day.

Chapter 8 Notes

1. Hayden, 2003.
2. Periplasmic flagella of the *Canaleparolina* spirochete are seen as dots in the cross-section figures in figure 6.2.
3. Fleck, 1979.
4. Duesberg, 1997.
5. UAB stands for Universitat Autónoma de Barcelona where this research is still under way.

From Kefir to Death

LYNN MARGULIS

The dairy drink well known to Russians and Scandinavians as "the champagne of the Caucasus Mountains" is a delicious, nutritious, live symbiotic food naturally selected during the past three thousand years by the long-lived Caucasians.

Death is the arrest of the self-maintaining processes we call metabolism, the cessation, in a given being, of the incessant chemical reassurance of life. Death, signaling the disintegration and dispersal of the former individual, was not present at the origin of life. Unlike humans, not all organisms age and die at the end of an interval. The aging and dying process itself evolved, and we now have an inkling of when and where. Aging and dying first appeared in certain of our microbial ancestors, small swimmers that were members of the huge group called protoctists (see chapter 4). Some two billion years ago, these ancestors evolved both sex by fertilization and death on cue. Not animals, not plants, not even fungi or bacteria, protoctists form a diverse—if obscure—group of aquatic beings, the smaller of which are called protists and can be seen only through a microscope. Amoebae, euglenas, ciliates, diatoms, red seaweeds, and all other algae, slime molds, and water molds are protoctists. Unfamiliar protoctists have strange names: foraminifera, heliozoa, ellobiopsids, and xenophyophores. An estimated 250,000 species exist today; most of them have been studied hardly at all. The vast majority that have ever lived are extinct.

Death is the loss of the individual's clear boundaries; in death, the self dissolves. But life in a different form goes on—as the fungi and bacteria of decay, or as a child or a grandchild who continues living. The self becomes moribund because of the disintegration of its metabolic processes, but metabolism itself is not lost. Any organism ceases to exist because of circumstances beyond its control: the ambience becomes too hot, too cold, or too dry for too long; a vicious predator attacks or poison gas abounds; food disappears or starvation sets in. The causes of death in photosynthetic bacteria, algae, and plants include too little light, lack of nitrogen, or scarcity of phosphorus. But death also occurs in fine weather independently of direct environmental action. This built-in death—for example, Indian corn stalks that die at the end of the season and healthy elephants that succumb at the end of a century—is programmed. Programmed death is the process by which microscopic protoctists—such as *Plasmodium* (the malarial parasite) or a slime mold mass—dry up and die. Death happens as, say, a butterfly or a lily flower made of many cells matures and then disintegrates in the normal course of development.

Programmed death occurs on many levels. Monthly, the uterine lining of menstruating women sheds as its dead cells (the menstrual blood) flow through the vagina. Each autumn, in deciduous trees and shrubs of the north temperate zone, rows of cells at the base of each leaf stem die. Without the death of this thin layer, cued by the shortening of day length, no leaf would fall. Using genetic-engineering techniques, investigators such as my colleague at the University of Massachusetts, Professor Lawrence Schwartz, can put certain "death genes" into laboratory-grown cells that are not programmed to die. The flaskful of potentially immortal cells, on receipt of this DNA, then die so suddenly that the precipitous cessation of their metabolism can be timed to the hour. The control cells that have not received the death genes live indefinitely. Menstrual blood, the dying leaf layer, the rapid self-destruction of the cells that receive the "death genes," and the slower but more frightening aging of our parents and ourselves are all examples of programmed death.

Unlike animals and plants that grow from embryos and die on schedule, all bacteria, most nucleated microscopic beings, the smaller protoctists, and fungi such as some molds and yeast remain eternally young. These inhabitants of the microcosm grow and reproduce without any need for sexual partners. At some point in evolution, meiotic sex—the kind of sex involving genders and fertilization—became correlated with an absolute requirement for programmed death. How did death evolve in these protoctist ancestors?

An elderly man may fertilize a middle-aged woman, but their child is always young. Sperm and egg merge to form the embryo, which becomes the fetus and then the infant. Whether the mother is thirteen or forty-three years old, the newborn infant begins life similarly immature. Programmed death happens to a body and its cells. By contrast, the renewed life of the embryo is the escape from this predictable kind of dying. Each generation restores the status quo ante, the microbial form of our ancestors. By a circuitous route, partners that fuse survive, whereas those that never enter sexual liaisons pass away.

Eventually, the ancestral microbes made germ cells, the egg and sperm, that frantically sought and found each other. Fusing, they restored youth. Animals, including people, engage in meiotic sex. They descended from microbes that underwent meiosis (cell divisions that reduce chromosome numbers by half) and sex (fertilization that doubles chromosome numbers).

Bacteria, fungi, and even many protoctists were—and are—reproducing individuals that lack sex lives like ours. They must reproduce without partners, but they never die unless they are killed. The inevitability of cell death and the mortality of the body is the price certain of our protoctist ancestors paid—and we pay still—for the meiotic sex their cells undergo.

Surprisingly, a nutritious and effervescent drink called kefir, popular in the Caucausus Mountains of southern Russia and Georgia, informs us about death. Even more remarkably, kefir also illustrates how the appearance of new species by symbiosis comes about. The word *kefir* (also spelled *kephyr*) applies both to the dairy drink and to the individual curds or grains that ferment milk to make the drink. These grains, like our protoctist ancestors but far more recently, evolved by symbiosis.

Abe Gomel, a Canadian businessman owns and manages Liberté (Liberty) dairy products just outside Montreal. He manufactures real kefir of the Georgian Caucausus as a small part of his line of products. Gomel and his diligent coworker, Ginette Beauchemin, descend daily to the basement vat room of his factory to inspect the heated growth of the thick, milky substance on its way to becoming commercial kefir. Like all good kefir makers, they know to transfer the most plump and thriving pellets at between nine and ten every morning, weekends included, into the freshest milk. Although nearly everyone who lives in Russia, Poland, or even Scandinavia drinks kefir, this "champagne yogurt" of the Caucasian peoples is still almost unknown in western Europe and the Americas. Abe Gomel and Ginette Beauchemin have been able to train only two other helpers, who must keep constant vigil over the two vats that are always running.

Legend says the prophet Muhammad gave the original kefir pellets to the Orthodox Christian peoples in the Caucasus, Georgia, near Mount Ebrus, with strict orders never to give them away. Nonetheless, secrets of preparation of the possibly life-extending "Muhammad pellets" have of course been shared. A growing kefir curd is an irregular spherical being. Looking like a large curd of cottage cheese, two centimeters in an irregular diameter, individual kefir pellets grow and metabolize milk sugars and proteins to make kefir the dairy drink. When active metabolism that assures individuality ceases, kefir curds dissolve and die without aging. Just as corn cobs in a field, active yeast in fermenting vats, or fish eggs in trout hatcheries must be tended, so kefir requires care. Dead corn seeds grow no stalks, dead yeast makes neither bread nor beer, and in the same way, kefir individuals after dying are not kefir. Comparable with damp but "inactive" yeast or decaying trout eggs, dead kefir curds teem with a kind of life that is something other than kefir: a smelly mush of irrelevant fungi and bacteria thriving and metabolizing, but no longer in integrated fashion, on corpses of what once were live individuals.

Like the protoctist ancestors of animals that evolved from symbioses among bacteria, kefir individuals evolved from the living together of some thirty different microbes, at least eleven of which are known from recent studies (table 9.1). These specific yeasts and bacteria must reproduce together—by coordinated cell division that does not involve fertilization or any other aspect of sex—to maintain the integrity of the unusual microbial individual that is the kefir curd. Symbiogenesis led to complex individuals that die (like kefir and most protoctists) before sexuality led to organisms that had to die (like elephants and us). A kefir individual, like any other, requires behavioral and metabolic reaffirmation.

During the course of brewing the yogurtlike beverage, people inadvertently bred for kefir individuals. In choosing the best "starter" to make the drink, villagers of the Caucasus "naturally selected," which means they encouraged the growth of certain populations and stopped the growth of others. These people inadvertently turned a loose confederation of

Table 9.1 Kefir: Components of Live Microbes	
Each individual Muhammed pellet is composed of:	
Kingdom Bacteria (Monera)=Prokaryotae	*Acetobacter aceti*
	Lactobacillus brevis
	Lactobacillus bulgaricus
	Lactobacillus casei
	Lactobacillus helveticus
	Leuconostoc mesenteroides
	Streptococcus lactis
Kingdom Fungi (yeasts, molds)	*Candida kefir*
	Kluyveromyces marxianus
	Saccharomyces cerevisiae, Torulaspora delbrueckii
and at least fifteen other kinds of unidentified but distinguishable microbes	

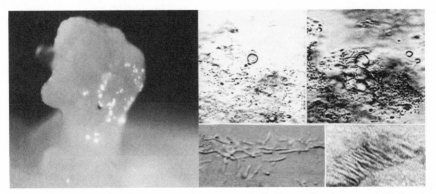

Figure 9.1 Left—the complex kefir "individual" live, magnified x2. Right—four views of kefir microbes inside the "Muhammed pellet," magnified x100.

microbes into well-formed populations of much larger individuals, each capable of death. In trying to satisfy their taste buds and stomachs, kefir-drinking Georgians are unaware that they have created a new form of life.

The minute beings making up live kefir grains can be seen with the light microscope in figure 9.1. Specific bacteria and fungi inextricably connected by chemical compounds, glycoproteins and carbohydrates, of their own making. These individuals are bounded by their own skin, so to speak. In healthy kefir, the bacterial and fungal components are organized into a curd, a covered structure that reproduces as a single entity. As one curd divides to make two, two become four, eight, sixteen, and so on. The reproducing kefir forms the liquid that after a week or so of growth becomes the dairy drink. If the relative quantities of its component microbes are skewed, the individual curd dies and sour mush results.

Kefir microbes are entirely integrated into the new being just as the former symbiotic bacteria that became components of nucleated cells are integrated. As they grow, kefir curds convert milk to both more of themselves and to the effervescent drink. Kefir can no more be made by the "right mix" of chemicals or microbes than can oak trees or elephants.

Kefir is a sparkling demonstration that the integration processes by which our cells evolved from bacteria still occur. Kefir also helps us recognize how the origin of a complex new individual preceded programmed death of the individual on an evolutionary time scale. Kefir instructs us, by its very existence, about how the tastes and choices of one species (ours) influence the evolution of others, the thirty intertwined microbes that became kefir. Although kefir is a complex individual, a product of interacting aggregates of both non-nucleated bacteria and nucleated fungi, it reproduces by direct growth and division. Sex has not evolved in it, and, relative to elephants and corn stalks, both of which develop from sexually produced embryos, kefir grains that undergo very little development display no sexuality. Yet when mistreated they die, and once dead, like any live individual, they never return to life as that same individual.

Knowing that symbionts become new organisms illuminates individuality and death. It is likely that the complex individual evolved in early protoctists similar to the way individual kefir pellets evolved. No sexual fusion was involved. Programmed aging and death apparently were profound evolutionary innovations, limited to the descendants of the sexual protoctists that became animals, fungi, and plants.

The development of death on schedule, the first of the "sexually transmitted diseases," coevolved with our peculiar form of sexuality. Sex is a process that kefir lacks now and always has done without. The privilege of sexual fusion—the two-parent "fertilization-meiosis" cycle of many protoctists, most fungi, and all plants and animals—is penalized by the imperative of death.

Welcome to the Machine

DORION SAGAN AND LYNN MARGULIS

In this painting by Christine Couture, we see a motley mix of the W. E. B. Du Bois tower library and the southwest dormitories of the University of Massachusetts–Amherst, Darwin's contemporary Samuel Butler on a computer screen, and parabasalid protoctists swimming about. We are, as Butler said, children not only of our parents, but of the plow, the spade, and the steam engine—and these days of the computer, the automobile, and the Internet.

So what's a warm, wet, furry creature like you doing in a place like this? Surrounded by cell phones, laptops, fax machines, and bar-code readers, we wise apes find ourselves wedged in the maw of a hectic technological revolution. Pierced and paged, wired and wireless, connected to streams of electrons and encased or enthralled by beating, bleating bits of metal, each of us is increasingly merged with devices of all kinds.

Benjamin Franklin, proud amateur of early explorations in electricity, might be at first incredulous, then fascinated, by the settlement of cyberspace and kindred developments. The futurists of the late nineteenth century could never in their wildest imaginations have extrapolated from that era to our own: too much would-be magic has already been made commonplace.

The United States and Japan, not Europe and China as in Franklin's era, set the standards for human experience today. Phantasms become palpable, the surreal incarnate: the projections of yesteryear—a landing pad on every suburban rooftop, a telephone in every Amazonian village—falter before images of a fossil spacecraft gleaming on a bleak moonscape, or a New Guinea tribesman, penis-board aloft, orienting himself by handheld satellite receiver.

An uncompromising metal buzz and an impersonal plastic crackle seem everywhere to replace the familiar voice, the parental hug, the childish cuddle. Have we finally cut the cord, or at least permanently

Figure 10.1 Intellectual ferment at the University of Massachusetts-Amherst.

transmuted it into fiber-optic cable? Have we departed the fertile fields and green gardens of our forebears on a one-way trip to a gleaming new technological paradise—or purgatory—distinct from all previous nature?

No. The machinate world that appears so new and unprecedented, so quintessentially and exclusively *H. sapiens*', is really not that at all. The glittering webs of the new communications, transportation, and genetic technologies are not simply cast over us by greedy corporations forced to sell what they overproduce. On the contrary, these new human-fostered technologies are in a direct line with the old. All arose from precedents— prehuman precedents—in an evolutionary and ecological context. Technology is a part of the human survival strategy, a prerequisite for human reproduction and population expansion; it has extended our ability to sense and manipulate the environment that supports us. It has been with us from the time long before we were human beings—that is, from before there even were any *Homo sapiens*.

For warm, wet, furry creatures like ourselves to bed down with electrical artifacts and electronic fabrications is, in short, entirely natural— entirely in keeping with life's ancient tendencies to expand, pollute, and complexify. It is our second nature and the nature of all of our ancestors.

Here we explore four propositions. First, that technology—the fabrication by living beings of useful objects and materials outside their bodies— is far more ancient than its tenure with modern humanity. Second, that life as a whole, not just human life, naturally incorporates its inanimate environment as it evolves. Third, that what begins as pollution in a growing population of thriving living organisms becomes the raw material for change as a species matures. And fourth, that machines and electronic devices are natural products of evolution, and are coevolving with us even as you read.

WELCOME TO THE TRIBE

Bone tools, fire-making flints, stone fishing weirs, and many other technological accoutrements coevolved with human families and groups of families before the beginning of modern humanness. Our chimpish
relatives—the funny one *Pan troglodytes* and the sexy one *Pan bonobo*,
who have more than 99.6 percent of their DNA in common with us—
fashion tools and communicate survival strategies among themselves. Any
separation of humanness from technology is delusional: from before the
beginning they were coupled. And no technique, no tool, no machine, no
sensing device was ever made by a single person alone. Technologies were
invented, refined, honed, and communicated by family, tribal, and even
larger groups. The technology of sticks and stones, of course, preceded
history's bronze, iron, and silica machines.

About three million years ago *Homo erectus*, *Homo ergaster*, and
other extinct people roamed the savannas and paddled the coastlines of
East Africa. As extended families and tribes, they absorbed deep knowledge of local settings. The details of vernal pond and spring source, the
timing of flowering and seed set, the course of fish migratory routes and
the hiding habits of small rodents were their objects of study. Those who
failed to learn or to share such knowledge died of starvation, thirst, or
treachery. Men or women who could not instantly recognize natural entities for what they were to them (drinkable, dangerous, toxic, or edible)
did not survive to see their children born and live to produce offspring.
Mental dichotomization, already established in our mute predecessors,
was a prerequisite for the survival of our jabbering *Homo* ancestors.
Ecological minutiae and biological detail had nothing to do with getting
into medical school or West Point; rather, memorization and comprehension of apparent trivia assured the supply of provisions—including
healing substances and arrow poisons—upon which our ancestors' pre-
carious lives depended. Very little was left to chance; abandonment of the
necessary to the random meant death. As populations enlarged, those
who best spoke and listened best imbibed the knowledge of the tribe.

Survival by learning—and its concomitant, the use of symbols—
became a central strategy of humankind. The summer band or winter
camp fellowship was always large enough to bring down the gazelle, find
the waterhole, deliver and tend the helpless infant, or navigate the croco-
dile-filled waters. No infant, adolescent, man, or woman, ever, in the history of mankind, lived in utter solitude; indeed, if isolated at an early

enough age an infant never becomes human at all. The mastery of technologies gave form to, and transmitted, knowledge: know-how. Flint-sparked fire followed the megafaunal hunt. Stone blades and clever traps brought down food on the hoof for well-led groups of jogging hunters, who celebrated their success by generating infants with potential like their own. Chattering babies matured to generate more—far more—prattling prodigies. These communicative humans consumed resources with increasing alacrity; their fabricating, practicing, teaching, and lovemaking wove a fabric of survival, and its patterns persisted in time and extended themselves in space.

Between six thousand and twelve thousand years ago, increasing numbers of humans were able to remove themselves from the rigors of the savanna hunt or the coastal rapids fishery. Social behavior in settled communities began to determine fecundity. But "hands-on" natural history still counted. Shepherds, farmers, potters, and basket weavers flourished even as stored grain allowed other means of making a livelihood to supplement ancient traditions of seed gathering and the chase. Religious ceremonies—drumbeat and wail—did just that: re-ligated, or tied once more, centrifugally dispersing bands into cohesive tribes, states, and nations.

O PIONEERS

The expansionist human species enjoys a population that has just now reached six billion souls. Humans dwell on every continent. At any given moment half a million people accompanied by pets, eyelash mites, and intestinal bacteria fly in airplanes overhead. Three million years after our origins, modern *Homo sapiens* continue to conform to the ecological type of the "pioneer species": those who move rapidly into new areas and grow rampantly, producing vast quantities of spores, seeds, or eggs.

Such species may inadvertently wreak havoc. Far from enhancing their own survival, pioneer species often subvert it. Pioneer grasses dry out the hospitable humus and convert it to dust; pioneer lichens convert receptive stone to soil. Chitons, a kind of flat mollusk with iron-magnetite teeth, chew South Pacific limestone islands at the waterline until they topple over and collapse.

Shortsighted pioneer species tend to be followed by more stable ones. Hardwood trees such as oak and beech replace fast-growing birch. Cyanobacterial communities replete with filamentous photosynthesizers stabilize the shorelines of sandy tropical islands. The hardwood forests and the solid cyanobacterial shores, known to ecologists as examples of "climax communities," may persist for thousands of millennia. Life goes on, and life that recycles its resources goes on longer. While pioneer species may so drastically alter their own means of livelihood that they destroy themselves, the climax communities that succeed them—and that may support whittled-down populations of the pioneers—persist longer in time even as they steady their immediate surrounds. They are more complex and more interconnected and generate greater rates of flow of more matter and energy; they are more "mature."

All species enjoy brief lifetimes, usually ten million years or fewer, relative to the vast stretch of geological time. For the most part disregardful of this truth, technological humankind continues to operate today as a pioneer species, moving through the habitats of others and converting non-human splendor to human convenience. Whether our fate will be the short life of a transient pioneer or a longer one as part of a planetary climax community is impossible to foretell. But short or long, pioneer or climax, the history of the human species will be inseparable from its technology.

Life—all of today's interacting thirty million species—naturally incorporates its inanimate environment. Life has fashioned, transported, made, and remade Earth's rocks, air, soil, and waters as it evolved from its bacterial origins over 3 billion years ago. No matter the details, all life requires energy, as either light or chemical reaction, and matter—some form of hydrogen, oxygen, sulfur, phosphorus, carbon, or nitrogen. Getting and spending this energy and matter, all beings alter their surroundings in species-specific ways.

Living beings, without exception, are made of cells: soft, pliable, watery, and vulnerable at their core. All take in nutrients of some kind and produce waste of a different sort. But the mode and speed of material transformation differs. From bacterium to shrub, from marine worm to social insect, life-forms reroute and reuse their waste. Their bodies chemically alter matter to produce hard substances: calcium ions from sea water, for example, combine with exhaled carbon dioxide to make the calcium-carbonate shells of the pearl oyster. Phosphoric substances combine with calcium in solution to form the calcium-phosphate tusks of the elephant. Excrement cemented with saliva constructs huge, air-conditioned, and humidified chambers that house tens of millions of tropical termites. In each case, soft cell material is surrounded and supported by hard parts of the body's own making.

Such home and body making represents the earliest of all technologies, for the biological production of hard minerals preceded by far the origin of apes, including human ones. Indeed, fabrication of hard mineral substance by living beings was in full swing long before any animal or plant evolved. Bacteria swim toward the bottoms of lakes, rivers, and seashores oriented by strings of magnets of their own making inside their bodies. Some marine protists among agglutinating *foraminifera*—huge but single-celled organisms that patch together their shells—choose round, black grains of sand from the immediate vicinity to make protective body cover from them. Some even fabricate towers. The foram crawls out and stands on the summit to peruse the menacing sea bottom that surrounds its homemade home base.

So often, in the history of life, what began as cast-off shell or anal exudate—as "excrement," "waste product," or "pollutant" in a growing population of thriving organisms—becomes a resource for change and expansion. Processes of recycling and reuse become increasingly refined and complex. Pioneer species die out, migrate, or settle. Climax species move in or increase their share of habitat. Their members engage in stable practices; their bodies often become habitat for other forms of life. The giant redwood, the coastal solanaceous tree *Lycium*, the saguaro cactus, and the Baja California boojum (*Fouquieria* or *Idria*) are not only individual trees in climax woodlands but food and shelter for birds, bats, rodents, flies, tree-hole algae, spiders, mites, termites, and basidiofungi. A legacy of life is to literally incorporate more and more of its environment into itself.

The longstanding tendency of life to co-opt its inanimate surroundings was documented in a most original way by the great Russian scientist Vladimir Ivan Vernadsky (1863–1945). Vernadsky recognized in his 1926 masterpiece *Biosfera (The Biosphere)* that the most important geological force is life. Two of the laws detailed by Vernadsky are that the number and kinds of chemical elements and compounds entering the cycling organization of living matter increase with time, and that as we move toward the present the pace of cycling increases.[1]

Human technological development is simply a recent example of Vernadsky's laws. Silica, now part of the human technological repertoire of particular importance in our computers, was in vigorous use three hundred million years ago by such marine protists as radiolarians and diatoms, who still make their shells of it. Synthetic isoprenoids, the rubberlike compounds now used in automobile tires, are but technological

Figure 10.2 Detail with printing press. Christine Couture.

incarnations of the rubber ooze from *Hevea* trees in the Colombian Amazon. Physicists have expanded the list of chemical elements circulating on Earth by creating new heavy radioactive elements like plutonium and seaborgium.

Long before humans, more and more chemicals of the universe were being sucked into living, proliferating life and its surroundings. Prehuman technologies—calcium shells, barium sulfate spines, phosphatic fecal pellets cemented into shelter—exemplify this tendency. Human technologies, especially complex contemporary technologies, extend this trend of nature.

As sentient individuals, we dearly love our gadgets, at least as long as they work well for us. We feel good as the rate of flow of energy and material goods increases around us; we feel irritated as it decreases. We perceive slowdown and cessation of the flow—brownout, system crash, meltdown—as boredom, malaise, even panic.

Yet it is artless to decry technology. Technology is part of nature; as Michael Heim writes, "Our hearts beat in the machines."[2] The lighting of fire, the binding of books, and the sewing of clothes are also forms of mechanical innovation, only by now such ancient features of the human landscape that they seem not machinate invaders but protective parts of

Figure 10.3 Detail with Samuel Butler on computer screen;
University of Massachussetts in background. Christine Couture.

ourselves. Whereas new technologies startle us, older ones—such as flush toilets—are so familiar they are noticed only by their absence.

Technology, in short, is an integral part of the ancient ecological cycles of procurement, removal, and reuse that appeared on Earth long before our ancestors turned human. And, of course, human technologies change even more quickly than prehuman ones did: they miniaturize, complexify, and prevail. Printing presses, punch-card computers, water pumps, fire-alarm systems, computing machines, hurricane-monitoring satellites—all instruments of design—began intrusively, being large and cumbersome in their early incarnations. Their descendants, by virtue of proximity, miniaturization, and pervasiveness, take on a more organic character.

BUTLER'S RAZOR

The genius of Charles Darwin (1809–1882) included his predilection for viewing humanity not as special and apart but as the product of a broader evolutionary process, and one that is still in progress. The intellectual legacy of Darwin allows us to perceive ourselves as a natural phenomenon, and the strength of his scientific worldview has proven extraordinarily powerful—even as it deflates our historical self-image.

Samuel Butler (1835–1902), considered "Darwin's most able critic" by the anthropologist Gregory Bateson's father William,[3] was fascinated by the evolution not only of organisms but also of the technologies they generate. The author of the Victorian classics *Erewhon* and *The Way of All Flesh*, Butler (1863) also published, under the pen name Cellarius, a sentence expressing with creative irony his ambiguous feelings toward the machine. "There is nothing," he wrote, "which our infatuated race would desire more than to see a fertile union between two steam engines." This extraordinary thinker also anticipated the Internet. In connection with the nineteenth-century invention of the telegraph—and while wryly predicting our enslavement by such mechanical servants—Butler, in 1863 (!) envisaged a day

> . . . when all men in all places without any loss of time are cognisant through their senses of all that they desire to be cognisant of in all other places, at a low rate of charge so that the back-country squatter may hear his wool sold in London and deal with the buyer himself—may sit in his own chair in a back country hut and hear the performance of Israel in Egypt at Exeter Hall—may taste an ice on the Rakaia [a New Zealand river] which he is paying for and receiving in the Italian opera house. . . . [This is] the grand annihilation of time and place which we are all striving for and which in one small part we have been permitted to see actually realised.[4]

The profound ability of human primates (with our domesticated plants, animals, and microbes) to work together for our common ends has had enormous survival value (relative, for example, to the loner orangutans and the less social chimps). Butler noted that engines are more efficient than draft animals at converting raw materials into human benefit and generally require less attention as well. Today, of course, the prowess of machines is greater by far than in Butler's time. Through us, machines manufacture more machines, which makes possible increased populations of people who cherish, utilize, alter, and generate still more, and more sophisticated, machines. Moreover, machines that take the place of physical strength are being augmented by those that replace mental power.

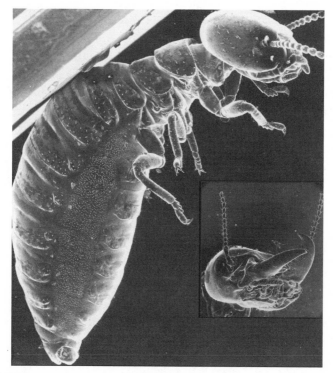

Figure 10.4 Termite: *Pterotermes occidentis* from the Sonoran Desert (inset: worker/soldier head).

From a biospheric point of view, machines are one of life's latest strategies for incorporating new elements and expanding life's role as a geological force. Like beehives, termite mounds, coral reefs, and other biological fabrications, machines—nourished by humans, themselves nourished by rice, wheat, cattle, and chickens—reproduce themselves. Agricultural contrivances such as tractors and harvesters produce food that encourages a vast and weedy growth of human populations; among these humans are agricultural engineers and entrepreneurs who design, develop, manufacture, and market yet more tractors and harvesters.

From the vantage point of the expansion of global life-forms, these machines are organelles—little organs—of a technological society. Just as the temperature- and humidity-regulated hives are crucial to the perpetuation of bees and termites, machines become now crucial to human survival. Indeed, the rate of evolution of machines today far exceeds that of people: machines grow exponentially, change rapidly, and reproduce the changed form more quickly than do the bodies of *Homo sapiens* or those of our best friend, *Canis familiaris*.

AVANT-GARDE THINKING

We fondly label the large, recent, expanding population of mammals of which we are members "evolutionarily advanced." We tend to equate recent evolutionary appearance, rapid change, and aggressive patterns of population growth with advancement. By these measures, however, our machines are more evolutionarily advanced than we are. They change form far more rapidly than we do: witness the automobile, the telephone, the photocopier, and the personal computer. And machines as a group can survive more extreme environments than can humans or our food plants and animals.

No mammal species, for example, unless it has evolved for millions of years in a watery environment, can survive underwater outside of a machine; manned and unmanned submarines function optimally beneath the sea. As extensions of ourselves in the accelerating rush of space travel, machines have left Earth's atmosphere and remained beyond it far longer than any person. Machines outperform people in such information functions as calculation and written communication. Machines have a range of energy at their disposal, such as nuclear fission, combustion, and photoelectric power; life's energy needs, by contrast, require precise forms of sunlight or specific inorganic oxidations and carbon-chemical reactions in water.

The love that we eager professors and students feel for our new laptops, software, color printers, Web access, guitar synthesizers, speaker connections, CD burners, backup disks, slide scanners, portable video projectors, and point-and-shoot cameras is a natural evolutionary impulse. So is the affection that our children feel for such devices: they love those recognizable aspects of their locale that feed, care for, and entertain them. The TV screen has tutored the current generation since infancy, and it is by machinate experience that today's students derive their connection to the material world. Surrounded as we are by the beige cases of electronic devices rather than the green and florid hues of plants, it is increasingly obvious that we, and especially our children, can no longer live what we take to be a civilized life without an elaborate tangle of electrically powered machinery.

Certain ecologically stable species, such as some subterranean tropical termites, display more technological prowess than others. *Heterotermes tenuis*, an Ecuadorian wood-eater, makes mounds, tunnels, and channels from its own feces and saliva; the ancestors of this species, as they outgrew their ranges, replaced wood-eating roaches that lived in logs. Analogously, human populations that extensively use machines outgrow those less dependent on them. Perhaps, emulating termite colonies, future human-machine communities will diligently recycle their own dead bodies, menstruum, sweat, and other exudates. Almost certainly our descendants will feed and move rapidly in ways that involve even larger quantities of food and numbers of machines. Cultural "excretions"—various discarded materials now labeled as sewage and pollution—will, we suspect, in accordance with Vernadsky's ideas, be brought effectively into the enlarging recycling system.

Unable, as yet, to self-reproduce, machines without humans have no more evolutionary staying power than shells without snails. Yet the trend toward human reliance on machines is obvious. Kidney-dialysis apparatus is a case in point. So is a dentures-wearing, pacemaker-equipped hominid whose overwintering survival depends on automobile access to refrigerated, processed, and mass-transported food. He is already a human-machine hybrid, flourishing at an evolutionary crossroads where the borders of organism and mechanism blur.

OUR MONITORS, OURSELVES

We begin to see ourselves and our technology more humbly within a universal schema of evolution of species with finite lifetimes. The leapfrogging spread of human-machine technologies over the global (perhaps eventually extraterrestrial) environment resembles the initial pandemonium of a pioneer species as it rushes across the tempting terrain of established climax communities. Aided by language and technological prowess and driven by our imperative to reproduce, we exploit new food and energy sources and create new exhalants, toxins, and population pressures.

But even as environmentalists lament the dehumanizing horrors of technology, even as naturalists pine for the green peace of virgin forests, human-fostered technology already reflects an evolutionary constraint. The rough-and-tumble mass production of the ocean- and atmosphere-contaminating industrial revolution already gives way to the smarter, subtler, more lifelike technologies of the information age.

Life on Earth is an evolving, interconnected system more and more efficiently using the sun's energy to recycle and recombine.

Mechanistic metaphor does life an injustice; all living beings, connected by common evolutionary history, are far more subtle and responsive than any machine. Life, unlike any clock or the motion of any planet, is no simple mechanical contraption: live beings, after all, are the creative source of all technology. Intelligent machine technologies flow outward from intelligent, sunlight-utilizing life; they are extensions of life's creativity. As ecologist/philosopher/sleight-of-hand-magician David Abram writes in his work on our sensuous experience, we humans have survived a period of haughty isolation in which we have allowed our ingenuity, alphabets, and linear thinking to alienate us from the natural world. But as Abram said to us in the 1980s, that period has also seen the "incubation" of our technologies.[5] Perhaps our evolutionary destiny is to use them not for our pioneering selves alone but for the prodigious expansion of all Earth's life.

Imagine the successful colonization of Mars, which will require not only fueled rockets with heat-resistant reentry tiles, sealed metallic containers, and human astronauts but also many kinds of food plants, protists, fungi, bacteria, and other animals as makers, keepers, and recyclers of the environment. This mental exercise demonstrates the future prospects for current technology—to perpetuate all kinds of biospheric life, not just urbanized machine-humans and our consumables.

As Butler insightfully observed, if and when a technology evolves long enough, it is no longer perceived as technology; it becomes integrated into the society of living organisms that use it, and of which it is now a part. Computer-orchestrated communities, Sun-supported space colonies, and underwater metropoli probably represent the frontiers for human-machine expansion. The arrogant habitat holocaust of today may cease; in its wake may evolve technologically nurtured habitats that re-bind, re-integrate, and re-merge us with nature.

But not necessarily the nature our nostalgia proposes. Our ancestors were East African apes, who, if we saw them today, would be thought to have escaped from a zoo. Our human-machine children and their humanoid offspring are likely to inhabit a transformed, machinate woodland, savanna, and shore—an Earth only superficially resembling the passing East African landscape and seacoast to which we long to return.

Chapter 10 Notes

1. Vernadsky, 1998.
2. Heim, 1993.
3. Bateson 1928.
4. Dyson, 1999, for a discussion in historical context of Samuel Butler's contribution.
5. Abram, 1996.

The Transhumans Are Coming

DORION SAGAN AND LYNN MARGULIS

We are already partly mechanical, connected by the tubes and
conduits of our external circulatory system of plumbing and
heating, our semi-invisible external nervous system of radio
waves and telecommunication stimulates us to emote and to act.
The robots have come quietly, gradually, without our awareness.
Robotic life and machine fossils—albeit sent by humans—
aready exist under the ocean, in the air, in orbit, and beyond on
the Moon, Titan, and Mars

Life may not progress, but it expands. Like an obsessive adult whose personality was formed in a forgotten childhood, humanity can be understood only as we make sense of our past. Earth life will have to evolve to
live on other planets, or even, perhaps, around other suns. And if as
humans we survive, we will certainly change, becoming part of the future
"supercosm"—the hypothetical continued expansion of life from Earth
into the solar system and beyond. The huge increase in area and resources
will unleash life's potential: the supercosm will be as different from Tokyo
as Washington D.C. is from a bacterium.

Human beings are peculiar parts of the biosphere, the place where life
dwells. The biosphere, the sum of life on Earth (the biota) and its surroundings, is part of us, and we have arisen from within it. As technologically dependent organisms, we have as much independence from the
biota as a virus has from the dividing cell in which it abides. Those twin
delusions of human grandeur—our natural superiority and scientific
objectivity—are conundra of projecting the techniques of human survival
into realms where they do not belong. The trial-and-error method of science, the forming and testing of hypotheses, and the rapid transmission of
science through culture are so similar to natural selection of hereditary
variants, on the one hand, and to survival and growth via bacterial
genetic transfer, on the other, that science can be considered as uncon-
sciously imitative and well within the scope of older biotic process.

Indulging the human mind's penchant for categorical choices and in keeping with our assignment to assess the future, we indulge in forecasts for humanity. Either there will be a catastrophic nuclear war that destroys our technologies or our technologies will control themselves so that, with machines, we begin reproducing in outer space. If the former occurs, people will vanish from the biosphere, global ecology will shift, and the biosphere will evolve in curious directions. It will not be a victory for humanity. However, if we survive our threat of nuclear war and become a multiplanet civilization, reproducing in outer space, this too will not necessarily be a victory for humanity. It will be a further expansion of the biosphere, a victory for the biota, for the nexus of all life, including machines.

The great ape *Homo* in his present state is a singularly technological creature. In truth, a human being may be thought of as an obligate technobe, a weak body entirely dependent on rapid harvesting of agricultural grasses and on milking, slaughtering, and packaging domesticated artiodactyls; on extraction of organic compounds, remnants of vast communities of photosynthesizers as fossil fuel oil from deep wells; on electromagnetic communication satellites, automobiles, and airplanes; in short, on machines.

Unfortunately for those who believe humanity is the apotheosis of life on Earth, the idea of reproducing machines is not a matter of scientific fantasy but a matter of fact in the present organization of the biosphere. Only the organic macromolecules DNA and RNA are capable of reproduction in the replicative sense: in one act of synthesis, they make complementary copies of themselves. All else—cells, boys, elephants, trees, McDonald's restaurants, and branches of the Chase Manhattan Bank— do not directly reproduce. Much molecular replication, cell growth, development, and construction is involved before two cells, two boys, two elephants, two trees, two McDonald's restaurants, and two bank branches appear in the biosphere where a single one was before.

Unfortunately, nature is not dichotomous in a way that matches our verbalizations. Nature does not conform to our definitions. Although there is an ineffable continuum between the living and the nonliving, we are beginning to understand the functions and organizations that are common to living entities. Living systems, from their smallest limits as wall-less bacterial cells to the entire surface of planet Earth, self-maintain. As living beings they are bounded systems—they retain their recognizable features, even while undergoing a dynamic interchange of parts.

We "modern humans" may never be the agents of the microcosm's expansion into space. Visual image processing in the form of eyes evolved many times; for example, it developed in dinoflagellate *Erythrodinium* protists, marine worms, mollusks (such as snails and squids), insects, and the ancestors to fish and mammals. Wings, likewise, evolved independently in insects, reptiles, birds, and bats: similar aerodynamic designs arose to meet the similar contingencies of the air. This tendency of organisms to evolve in similar directions despite the fact that they have different recent ancestors is called convergence. Convergence suggests that many kinds of beings will expand into space, just as many kinds have moved onto dry land and into the atmosphere. But like the first lungfishes, which came out of water but never evolved into the ancestors of land animals, these early flirtations with space may never be consummated by continued life there. The presence of nervous systems and community behavior in many sorts of animals suggests that if we people and our "urban" associates fail, other life-forms will evolve to cart the primordial microcosm into space. If human beings become extinct—or if, like the horseshoe crab or lungfish, we just happily remain in our present habitats—the biota may, for a time, remain confined to Earth. But remember it took humans (*Homo*) only a few million years to evolve (from *Australopithecus*). Even if all anthropoids—all humans, monkeys, and apes—became extinct, the microcosm would still abound in those assets (for example, nervous systems and manipulative appendages) that were leveraged into intelligence and technology in the first place. Given time to evolve in the absence of people, the descendants of raccoons—clever, nocturnal mammals with good manual coordination—could start their own space program. Sooner or later the biosphere is likely to expand beyond the cradle of this third planet.

It is an illuminating peculiarity of evolution that explosive geological events in the past have never led to the total destruction of the biosphere. Indeed, like an artist whose misery catalyzes beautiful works of art, catastrophe seems to have immediately preceded major evolutionary innovation.

Life on Earth answers threats, injuries, and losses with innovations, growth, and reproduction. The disastrous loss of hydrogen gas (H_2) from the gravitational field of Earth led to one of the greatest evolutionary successes of all time: the use not of H_2 but of H_2O (water) in photosynthesis. But this substitution of a necessary ingredient also led to a devastating pollution crisis. The accumulation of oxygen gas in the atmosphere, a gas originally toxic to the vast majority of organisms, permanently changed the planet. The oxygen crisis that began only two billion years ago prompted the evolution of respiring bacteria. These microbes that used oxygen to derive biochemical energy more efficiently than ever before eventually took over most of the world. Some of the oxygen-breathing bacteria became symbiotic, merging with different (oxygen-eschewing) bacteria to form nucleated cells, which, after becoming sexual, evolved into fungi, plants, and animals. (See chapter 13 for how sex may have evolved.)

The most severe mass extinctions the world has ever known, at the Permo-Triassic boundary 248 million years ago, were rapidly followed by the rise of mammals, with their sharp eyes and large, receptive brains. The Cretaceous catastrophe, including the disappearance of the dinosaurs 65 million years ago, cleared the way for the development of the first primates, whose intricate hand-eye coordination led to technology. World War II ushered in radar, nuclear weapons, and the electronic age. And the holocaust of Hiroshima and Nagasaki over sixty years ago decimated Japanese industry and culture, unwittingly clearing the way for a new beginning in the form of the rising red sun of the Japanese information empire.

With each crisis the biosphere seems to take one step backward and two steps forward—the two steps forward being an evolutionary solution that surmounts the boundaries of the original problem. Not only meeting but transcending challenges confirms the resilience of the biota. The denizens of the biosphere habitually recover from tragedies with renewed vigor. Nuclear conflagration in our hemisphere here in the north would kill hundreds of millions of human beings. But it would not be the end of all life on Earth. Far from it. As heartless as it sounds, a human Armageddon might prepare the biosphere for less self-centered forms of living matter. As different from us as we are from dinosaurs, such future beings may have evolved through matter, life, and consciousness to a new superordinate stage of organization. They might consider human beings as impressive as we do iguanas.

Such a vision offers only metaphysical consolation. Barring direct fatal impact by an atomic weapon, only ten micrograms (that is, ten millionths of a gram) of radioactive fallout—the debris that explodes into the stratosphere, blows in the wind, and later settles down—is needed to kill a person. Current estimates put Russian and U.S. nuclear bomb arsenals at ten thousand megaton bombs apiece. As the late inventor Buckminster Fuller showed by dropping tiddledywinks on a giant map spread across the ballroom floor of the New York City Sheraton Hotel, five thousand bombs released at random on the globe would paralyze all the major cities. And given present arsenals, a full-scale nuclear war can be expected to deplete from 30 to 60 percent of the ozone of the stratosphere. The dust and smoke of city fires would rise up and surround Earth, first burning it but later leading to a severe drop in worldwide mean temperature.

Radiation could also accelerate worldwide plagues of AIDS-like and other diseases with compromising effects on the human immune system. Yet the health and stability of the microcosm might even be strengthened. The increase in radiation-induced mutations wouldn't change microbial evolution because a huge reserve of radiation-resistant mutants to supply the evolutionary process has always been present. *Micrococcus radiodurans* (now called *Deinococcus*), for example, has been found living in the water used to cool nuclear reactors. Nor would the destruction of the ozone layer, permitting entry of torrents of ultraviolet radiation, ruin the microbial underlayer. Indeed, it would probably augment it, because radiation stimulates the bacterial transfer of genes.

The accelerated nature of evolution in general and cultural evolution in particular makes it impossible to predict future evolutionary change, especially at long range. If we simply extrapolate current trends, we arrive not at the future but at a caricature of the present. When the telephone was invented, in an anecdote told by science-fiction writer Arthur C. Clarke, it was predicted that, in the not-too-distant future, every city and town might have one of its own. When helicopters first appeared, on the other hand, there were commentators who saw the day when every suburban household would park its private twirly vehicle near its automobile in a heliport-garage. Respectable scientists, writing in technical journals with full citations to the professional literature and mathematical equations, predicted that the surface of the Moon was covered with commercially exploitable levels of oil. Some stated that desert lichens, seasonally turning green, grew to nearly cover an entire hemisphere of the planet Mars with each moist summer. Other scientists predicted thick dust layers would so impede a lunar landing that explorations of the Moon would be impossible and ought not be attempted. We certainly do not pretend to have knowledge of the future, but we do prefer to contemplate possibilities based on an awareness of our long-term past.

Beyond short-term technological fads are the long-term trends of life—extinction, expansion, symbiosis—that seem universal. We, the species *Homo sapiens*, will reach extinction, with or without a nuclear war. We may, like ichthyosaurs and seed ferns, leave the annals of Earth history

without an heir, or we may, like choanomastigote protists, australopithecines, and *Homo erectus* mammals (the respective ancestors of sponges
and of us), evolve into distinct new species.

No matter what our progeny evolves or devolves into, however, if it
remains on Earth eventually it will be scorched alive. By an astronomical
reckoning, our Sun has a total life span of only about ten billion years.
After all the Sun's primary hydrogen burns up as fuel, nuclear reactions
that convert lighter to heavier atoms are expected to take over. As the
radiating Sun expands into a red giant, our dying star will shine as it has
never shone before. The luminous body is expected to generate such
immense heat that oceans will boil and evaporate.

As Earth burns up, its oceans boiled to steam by the final outbursts of
a waning Sun, only living forms that have wandered beyond this home
planet or have protected themselves in some way will be salvaged. As is
the wont of life, the habitats of life's predecessors will be brought into the
homes of life's future. The insertion of past dwelling places into new ones
is an intense sort of conservatism, a deep-rooted refusal to change that is
observed in tropical beehives, naked mole rat tunnels, and Russian émigré
populations at Villefranche-sur-mer (Mediterranean France) or San
Francisco's valleys. Such a monomania for preservation may be just what
is needed to rescue future organisms, and with them life itself, from the
fate of an exploding Sun.

We already see hints that the boundaries of life, human and cockroach,
pet and grain, are expanding. Populations, industries, universities, and
suburbs have rapidly grown, but none has grown indefinitely without
causing severe resource depletion and environmental transformation.
Natural selection, which simply refers to different rates of survival among
growing reproducing entities, whether of monkeys or of McDonald's
restaurants, can prune or frighten. Population growth is limited, and populations are beyond good and evil. All living forms grow in response to
the availability of space, food, and water. When too numerous, all organisms either perish or transcend themselves.

Organisms that transcend themselves always find new ways to procure
Lebensraum (room to live), carbon, energy, and water. All this expanding
beingness produces new wastes and new needs for space, food, and water.
The increasingly abundant production of new wastes stresses those that
made it. Life itself expands without much remorse and creates its own
new problems; life forces new solutions. One can imagine an example:
Pollution might be created by the venting of new chemicals in the outer

solar system as part of a program of resource acquisition by future corporations. Such toxic wastes might even reach Earth. On Earth, new microbes able to tolerate or make use of such wastes might be forced to evolve. This, in turn, would establish a living partnership that stretched millions of miles, from Earth to the moons of Saturn.

To grasp the potential of life in the future, we must look at life in the past. The dramatic evolution of humans cannot be separated from the coevolution of our microbial ancestors, the bacteria that constructed our cells and those of our food species of plants and animals. In coevolution, over thousands of years partners change genetically. Inherited partnerships evolve together as new proteins and developmental patterns emerge. When ultimately the partners totally depend on each other, they become larger new entities. No longer is it valid to consider them independent or separate individuals. The agricultural grain *Zea mays* (corn) provides one striking example of coevolution. Corn has evolved in a few hundred human lifetimes, during the past six thousand years. Corn on the cob no longer withers naturally as do the seed-releasing flowers of the teosinte grasses from which modern corn evolved. Corn now must have its thick husk removed by human hands in each and every generation. Similarly, cows must be milked, and chickens fed. Now the reproduction of our food sources—corn, cattle, and chickens—is tied to our own. Corn cannot complete its life cycle without people; these organisms form a part of us. Once an inconspicuous, self-sufficient grass on the Mexican plateau, the plant teosinte has been selected by hungry peoples and has been grown for larger and larger kernels. It has become a major staple for humanity. Like the electric wires in the elevators of Manhattan and Los Angeles, the luxury of yesterday has become the necessity of today.

The prodigious increase in the human population has depended on plants, and probably will continue to depend on them and their bacteria-derived chloroplasts if we are to move into space. It took a thousand hectares during the last interglacial period to support a single Old Stone Age hunter. Over ten thousand times less agricultural land is required to support a modern Japanese rice farmer. Thus for every hunter-gatherer that once roamed the island of Honshu, over ten thousand inhabitants in a Tokyo suburb may thrive. Like the cells of the microcosm before us, human beings must coevolve with plants, animals, and microbes. Eventually we will probably aggregate into cohesive, technology-supported communities that are far more tightly organized than simple or extended families, or even nation-states or the governments and subjects of superpowers.

Because new symbioses tend to form during evolution and any organism is always a member of a community of different species, no single life-form or member of one species alone could ever colonize space. Humans seem well suited to help disperse Earth-based biota, and they may occupy a prominent place in the supercosm—just as mitochondria (former oxygen-using bacteria now permanently inside cells of plants and animals) helped the mosses and ferns, amphibians, and anthropoids settle dryer land. But for us humans to play the prominent role in the expansion of life into space, we must learn from the successful communities of the microcosm. We must move more rapidly from antagonism to coexistence. We need to treat the members of species whose health is of interest to us as fairly as a small farmer does his egg-laying chickens and milk cows. Unlike poaching rare animals for their pelts, garishly displaying horned heads over a mantelpiece, shooting birds for sport, or bulldozing rain forests, such fair treatment means cohabiting the plains and forests with our planetmates. Contrary to his hunting ancestors, the small farmer of today does not destroy a chicken or cow for a single feast but nourishes populations of his animals, consuming their milk and eggs.

This sort of change from killing nearby organisms for food to helping them live while eating their dispensable parts is a mark of species maturity. It is why agriculture, in which grains and vegetables are eaten but some of their seeds are always stored, is a more effective strategy than the simple plant gathering. The trip from greedy gluttony, from instant satisfaction, to long-term mutualism has been made many times in the microcosm. Indeed, it does not even take foresight or intelligence to make it: the brutal destroyers destroy themselves, while those who interact more successfully inherit the living world.

Even with an understanding of our origins, our view of our future blurs the further we look. But as the visionary poet William Blake wrote, "What is now proved true, was once only imagined." There are many imaginable ways by which people might evolve into a species distinct from *Homo sapiens*. The simplest would be not only by the accumulation of random mutations but by sexual recombination of preexisting genes. Although all human beings belong to the same species, population extremes may be noted. A Pygmy woman, for instance, may not be able to give a Watusi man a baby because her pelvis is too small. This example illustrates the natural variety present in any species, which may, over time, give rise to divergent species unable to interbreed because of outward changes resulting from inner ones: altered symbionts, different behaviors, rearrangements of chromosomes, changes in mitochondrial genes, duplications of nucleotide sequences in the DNA, or others.

But cells can now be fused in forced fertilization, and the simple accumulation of vast numbers of changes in DNA base pairs can now be engineered. The genetic "writings" of future biotechnologists ultimately may be new organisms. The use of sets of bacterial genes—or at least the funding for such use—has already become commonplace. Through biotechnology those pieces of DNA called plasmids are inserted into bacteria and thus quickly replicated. Genes coding for proteins, even human proteins, may be replicated via association with plasmids.

The fascinating question of direct intervention in human evolution is approachable from several separate fronts, including traditional natural selection (deforestation, animal and plant breeding) as well as newer techniques: biotechnology, computers, and robotics. Given that evolution accelerates, it must be only a matter of time before these approaches converge. Geologically speaking, we refer to exceedingly brief time periods, even within our children's lifetimes.

Computer science has been one of the most rapidly growing fields in the history of technology. From vacuum tubes to transistors and semiconductors, the information-handling elements of computers have miniaturized tens of thousands of times in only several decades. Their switching speed, the time required to switch on to off in a binary code, has decreased from twenty to a billion times per second.

As computerized records, books, and other devices become commonplace because the raw, siliceous, and miniaturized components of computers are so inexpensive, society will transform. The trend for money to become increasingly electronic will continue. Education will become easier as teaching gadgets enter the market. Beyond the "paperless office," there will occur what the computer expert Christopher Evans called "the death of the printed word." Traditional printed books will become as extravagant—and as expensive—to people of the future as first editions or hand-printed manuscripts seem to us. Books will appear to be immensely laborious undertakings. Each bulky mass of ink-spotted paper will take on the antiquated aspect of the Mainz Bible of Johannes Gutenberg. Because the complex nature of future societies is bound to be dependent on and monitored by computer intelligence, social movements, financial transactions, and exploratory discoveries will be recorded in machine memories. Because retrieval of computer-stored events will be far more faithful than movie "re-creations" or historical novels, it will be possible to relive history. Through technology, life's ancient ability to pre-serve the past in the present, its mnemonic fidelity, will vastly improve. This memory phenomenon, aided by cinema, written history, electromagnetic records, and other computer technology, is still accelerating.

Because silicon chips with thousands of bits of memory can pass through the eye of a needle today, microprocessors—tiny computers—are now lightweight enough to insert into machines, making them robots. Robots have great potential for the future. In 1976 the robotic part of the *Viking* spacecraft performed a task no human being could have done: landing on the ultraviolet-light-bombarded, frozen, and suffocating surface of the red planet, it stretched its mechanical arm, drew in a sample, and analyzed the dry and oxidized Martian regolith. Other robots are more mundane. Metal robots with many arms fasten tires to cars with a productivity rate that far exceeds that of their human counterparts. The assembly line itself is becoming assembled. Robots in Japan, for example, make parts for other robots.

As computers and machines come together in the new field of robotics, so robotics and bacteria may ultimately unite in the so-called biochip. Based not on silicon but on complex organic compounds, the biochip becomes an organic computer. Manufactured molecules, like photosynthesizing plants, would of course exchange energy and heat with their surroundings. Energy would be converted not into cell material but into information. The possibilities inherent in such a development are awesome. "Living" computers could trade millions of hydrogen atoms per second and perhaps be integrated into conscious organisms. At this distance in the future the imagination is overwhelmed. The outcome of information exchange between computer, robotic, and biological technologies is not foreseeable. The most outlandish predictions, in retrospect, will seem naive.

What are possible fates of *Homo sapiens* in the next centuries? Let's explore two of many. As we have seen, the nucleated cells of all animals, fungi, and plants contain genes packaged as chromosomes. Species are known to evolve by several means, including chromosomal rearrangements, the accumulations of mutations in DNA, and symbiosis. Chromosomes undergoing heritable changes can cause jumps in evolution larger than those caused by nucleotide base-pair mutations. Symbiotic leaps can, in a few generations, establish new species. Such modes of variation should operate on populations of people. Abrupt chromosomal changes, such as those involved in karyotypic fissioning, have led to many new species of mammals. Karyotypic fissioning is the name of a process in which chromosomes break apart at their centers. Many species of Cenozoic mammals, compared with their ancestors, show half chromosomes, broken at their centers. Dr. Neil Todd[1] has shown how karyotypic fissioning has led to the evolution of dogs from wolves, pigs from boars, and even the humanlike apes from their apish ancestors.[2] Combined with incest, karyotypic fissioning, in principle, may lead to new species of humans. The conquerors of the supercosm, if they are our descendants, or at least the descendants of some of us, are likely to have even more fissioned chromosomes than we do now and to have new traits, such as the ability to move easily, grow, and reproduce under decreased gravity.

Future humans may even be green, a product of symbiosis. An example of such a symbiotically produced species of human is *Homo photosyntheticus*, the imaginary cure to the heroin problem suggested by the algae expert Ryan Drum. *Homo photosyntheticus*, he claims, would be descendants of heroin addicts whose heads had been shaved and injected with a thin layer of algae. Strung out under the lights, such green hominids, who would not have to be addicts, would be fed by their internal resources and, as Drum suggests, far less of a social burden.

Evolution has already witnessed nutritional alliances between hungry organisms and sunlit, self-sufficient bacteria or algae. *Mastigias*, a Pacific Ocean medusoid, a peaceful coelenterate of the man-of-war type, helps its photosynthetic partners by swimming toward the areas of most intense light. They, in return, keep it well fed. This could happen to our *Homo photosyntheticus*, a sort of ultimate vegetarian who no longer eats but lives on internally produced food from his scalp algae. Our *Homo photosyntheticus* descendants might, with time, tend to lose their mouths, becoming translucent, slothish, and sedentary.

Symbiotic algae of *Homo photosyntheticus* might eventually find their way to the human germ cells. They would first invade testes and from there enter sperm cells as they are made. (This is hardly outrageous: insect bacterial symbionts are known to do exactly this. Some enter sperm, and some are transmitted to the next generation via eggs.) Accompanying the sperm during mating, and maybe even entering women's eggs, the algae—like a benevolent venereal disease—could ensure their survival in the warm, moist tissues of humans.

In the final stages of this eerie scenario, we envision groups of *Homo photosyntheticus* lounging in dense masses upon the orbiting beaches of the future, idly fingering green seaweeds and broken mollusk shells.

Electronically connected to their bank accounts, they would have no incentive ever to hurry.

We have suggested two possible paths of the evolution of humans. They are fanciful, perhaps, but the lessons of the past tell us that even if our details are absurd, dramatic changes are inevitable. We can think of other peculiar possibilities. One is cybersymbiosis, the evolution of parts of human beings in future life-forms. People in this scenario are as crucial to the development of the supercosm as bacterial interaction was to the macrocosm. If we do transcend the fate of mammalian extinction and survive in an altered form, we may persevere not as individuals but as remnants. We can imagine ourselves as future forms of prosthetically pared people—with perhaps only our delicately dissected nervous systems attached to electronically driven plastic limbs and levers—lending decision-making power to the maintenance functions of reproducing spacecraft.

Chapter 11 Notes

1. Todd, 1970. (For further details, see Kolnicki 1999 and 2000 in the *Readings* that begin on page 238)

2. The importance of centromere-kinetochore reproduction in the karyotypic fissioning process for mammal evolution is diagramed in Kolnicki 1999 and 2000. As a mode of speciating, the process has been illustrated for old world monkeys and apes (Giusto and Margulis, 1981), carnivores and artiodactyls (Todd, 1970) and lemurs (Kolnicki, 2000).

Alien Enlightenment: Michael Persinger and the Neuropsychology of God

DORION SAGAN

A Canadian neuroscientist suggests that aliens may be out there but our experience of them comes from in here—from electromagnetic stimulation of the brain.

It was very curious. The night before I went to work on this piece about Michael Persinger's theory of magnetism tweaking the brain to experience alien abductions, I found myself at a book party. It was in Bedford, Massachusetts, in honor of Ernst Mayr, a lucid ninety-eight years old and perhaps the top evolutionist in the world. A spattering of the scientific glitterati were here–editors, scientists, journalists, and one eight-year-old girl, feet dangling over the edge of the stage. Amid the question-and-answer period of the festivities, which consisted mostly of an impromptu symposium about the role of symbiosis in evolution, she wondered aloud why we, and more to the point, she, were there. I agreed with her. Also in attendance was Janet Williams, mother of Paul Williams, author of the 1960s cult classic *Das Energi* and former executor of Philip K. Dick's literary estate. No one has more expertly massaged the metaphysical subtleties of ultimate reality into schlock pulp than Dick, so I must mention him.

Later, I found myself across the table from Mayr and his daughter Christa. Why, I asked him, did he think the chances so poor for life to exist in space? He responded that the question had to do not with life but with intelligent life. Of the several billion species that had existed in Earth's history, there was only one that was perhaps intelligent: us. He added that most who studied the matter were not even biologists. When he spoke with him even E. O. Wilson, the great Harvard biologist, had to admit, said Mayr, that the chances of life existing in space were vastly improbable. "But wouldn't it be great if we received a message," Wilson mused.

I presented my epigram that the search for extraterrestrial intelligence was a replacement for religion in a secular age. I shared this quip years ago with my father, who went on, in the book *Demon-Haunted World*, dedicated to my son, to postulate that reports of abductions were often repressed memories of early childhood; even without any objective sexual component, the polymorphously perverse infant, a synesthetic mass of sensations, would be snatched from its crib and handled by that sublunar prototype of gods and aliens: parents.

Christa remarked that each age had its weird belief system. The previous century it had been ghosts taking time out from their heavy schedule in the afterworld to meet with a medium and the mourning in séances; now it was high-tech aliens from the heavens. Bill Frucht, an editor from Perseus Books, put down his wine. "See, I disagree," he said. "Even finding slime on Europa would be a great discovery."

"Even the stars could be alive," I said, "but we wouldn't know it." I stole this idea from Californian poet Robinson Jeffers, and made the mistake of admitting it. "We can't even communicate with other species on Earth," I said. "How can we expect to understand aliens? And what about whales?" I pressed. "Couldn't they be more intelligent than we are?"

"The question shows that you don't understand the question," Mayr said. "How can there be intelligent life in space? There is not even intelligent life on Earth." It figured: although he had been slated to take the no-extraterrestrial-life side of a debate with my father, Carl Sagan, in 1996 before Dad died, now he sounded like him (one of my father's early talks, for a primarily African American audience in Alabama, was "Is there intelligent life on Earth?").

The next morning I crept under the covers with my spouselike object (her term). She accused me of just wanting to have sex with her. Nonetheless, the warmth of her body eased my anxiety about encroaching deadlines and her imminent departure. Soon I was asleep. The presence of her seemed to activate my mind. I had strange dreams. I was applying shampoo to my hair, but I was not in the shower. When she awoke and I told her about it she started: she had had the same dream. I was applying shampoo to my hair, outside the shower. In her dream the shampoo was a pink powder. It could be the exact same dream, given that mine was in black and white. Strange. Could the presence of her rare head next to my soft skull have literally turned on something in my brain? Was this connection electromagnetically mediated? Was it a "Persinger moment"?

Perhaps ESP is electromagnetic, some kind of invisible pulse. But the abbreviation is oxymoronic, a contradiction in terms: if we can perceive it, it is not extrasensory. Whatever the channel, it is sensory, not extrasensory, perception, even if stimuli bypass the retina and balance organ and proceed directly to the source, the brain.

Now I had to speak with Persinger. When I tracked him down he told me it was remarkably easy to produce sensations of a nonexistent being in one's midst. In one case 60-hertz waves from the choppy electromagnetic field of a digital clock correlated with a young woman's sensation of an alien in her bedroom; the intruder took his dastardly desires with him as soon as the clock was removed. One house, an electromagnetic maze of interacting fields, sported multiple computers and stereos in close proximity. Needless to say, it was haunted.

"It is not just the intensity," explained Persinger. "It is the complexity of the temporal structure. Think of sound. Someone could hammer away, producing a 100-hertz sound wave, but it has no meaning; it is just background noise. But if someone whispers 'help me,' immediately they have your attention. Such a sonic pattern has information. It is the same with EM fields and the brain."

I asked Persinger whether he had read Philip K. Dick. He had not; he had too many students to oversee, he was too much of a laboratory animal. But their views seemed similar. Both operated under the notion that everything from "the outside world" must go through the brain. The outside world is on the inside.

This Möbius striptease of an epistemology may be more realistic than we suppose. We like to think the seam between what is outside and inside our heads as being pristine. Perhaps it is not. This is discomfiting at first but leads to a sort of satori. If brains produce electromagnetic fields, perhaps under certain conditions we can pick them up through the thickness of the skull. Extrasensory perception, insofar as it exists, and even more so extraterrestrial abduction, insofar as it does not exist, have an obvious middleman: the brain. The brain is the great mediator of the senses—the place where they come together. I therefore hereby suggest we speak of "infrasensory" perception—for the brain is not beyond the senses ("extra") but beneath them ("infra"), making, well, sense of them. Similarly I suggest that extraterrestrials be called not ETs but ICs—"intracerebrals."

As P. K. D. said, "The stars are inside us," a statement both mystical (suggesting we perceive space inside out) and literal, since what registers are not the stars themselves but electrochemical traces stimulating changes along our retina and optic nerve and into the visual parts of our information-processing brains.

Persinger's work can be divided into two categories that overlap: a minor investigation of geomagnetic phenomena and a major quest to find the specific electromagnetic algorithms underlying specific experiences in the human brain. An understanding of how these overlapping categories work allows us to provisionally explain a host of psychic phenomena, from out-of-body and near-death experiences to being held in the arms of a loving God to close encounters with putative aliens. Lest this should molest those who hold dear space-going sylphs armed with probes and a taste for the odd unsavory medical procedure, I want to reassure you that Persinger's scientific demystifications open a Pandora's box of more exotic vastations—specifically, the burgeoning of a brain tech, replete with electromagnetic helmets that make the drug-soaked 1960s look like a sip of wine in the park. Indeed, that is what Dick, in his alluring mix of science fiction and metaphysics, in his stories of brain mood changers and memory implants, of Jesus's image projected onboard distant starships by deviant aliens, was getting at.

Born in 1945 in Jacksonville, Florida, Persinger had his eyes on the interdisciplinary prize for a long time. He grew up mainly in Virginia, Maryland, and Wisconsin and majored in "psychochemistry" at the University of Wisconsin in Madison, from which he graduated in 1967, the Summer of Love. With a scientific mind unwilling to shrink from the raw, fascinating data of cultural upheaval in the Dionysian sixties, Persinger was drawn to the links between society, the physical sciences, and the human brain. He received his MA in physiological psychology from the University of Tennessee, and his PhD from the University of Manitoba in Canada in 1971. Since then he has been a professor at Laurentian University in Sudbury, Ontario, a nickel-mining town. Drawn to discover the reality behind reports of psychic and alien phenomena that were anathema to traditional science, Persinger compiled powerful interdisciplinary tools: statistics (to move from anecdote to correlation), geophysics ("because it is a central focus of the physical sciences"), and neuroscience ("a central focus for the emerging biosocial sciences"). Magnetic fields offered themselves to him as "one of the few stimuli that evoke changes across all levels of scientific discourse." At Laurentian he inaugurated the behavioral neuroscience program, which recruited students with less than top grades but adept at synthesis and receptive to Persinger's mix of chemistry, biology, and psychology. Meanwhile Persinger became a registered psychologist, a clinician specializing in mild head traumas; he also explored the commercial possibility of treating depression and other ailments with magnetic fields.

It had long been known that epileptics are prone to mystical visions during seizures. Music and ritual have been shown to affect the limbic system (located in the temporal lobes on the sides of the brain), and surgical stimulation of this area of the brain—which we share with other mammals, and which is deeply connected with emotions—can bring on what Henry James Sr., father of the novelist Henry James Jr. and his great psychologist brother William, called his "Swedenborgian vastation" (after church founder Emanuel Swedenborg). Russian novelist Fyodor Dostoyevsky wrote of touching God during epileptic fits; Joan of Arc was an epileptic, as were a disproportionate number of mystics. The epilepsy connection, Persinger told me, suggested that the amygdala, known to color experiences with meaningfulness, may have been electrically stimulated by magnetic bursts during seizures. Could the beneficial aspects of such seizures be reliably reproduced?

After Persinger's 1987 book was published, he began a systematic study of complex electromagnetic fields.[1] Persinger's friend Stanley Koren, "the technical genius," built the equipment, the Koren or Persinger helmet. I call them "happy helmets," as it amuses me to hear stories about them. After experimenting with different patterns Persinger found that a weak magnetic field rotating counterclockwise around the temporal lobes caused four out of five people to feel a distinct and mysterious presence. The 1-microtesla field is not intense—it is emitted at the same frequency as your computer screen. But it is modulated: the experience is prepared for by twenty minutes of exposure over the temporal parietal lobe of the right hemisphere only; this is followed by four seconds of rapid so-called burst firing alternating with silence. Then, more often than not, an "ego-intruder" is felt—the presence of someone or something else.

What we interpret this presence to be seems to depend on our cultural predispositions. In Victorian days it was more likely to be an angel or the Virgin; these days aliens appear to have the upper hand. The extra stimulation to the right hemisphere is important. Calling the self "the last big challenge in neuroscience," Persinger argues it—the ego, the "I"—is largely a linguistic construction of the left hemisphere. "You don't experience the right hemisphere as you," he says. Nonetheless, these brain processes are right there, sharing the skull. Without chitchat from the left side of the brain, we can feel communion with "the great cosmic goop."

Alternatively, stimulation of the right side of the brain can in theory spur misrecognition by our literal-minded left brain, scaring us of our own electromagnetic shadow, into thinking someone is present when no one is there.

Persinger stresses that his research in no way is meant to demean anyone's religious/mystical experience. "We all know somebody who has dramatically changed as a result of a mystical or religious experience," he told me. "Their depression vanishes; they become a new person. What if we could cut out the cosmological interpretation?"

The notion of producing the reliable equivalent of satori, the repeatable experience of enlightenment, is at the forefront of Persinger's agenda. There is the possibility that electromagnetic (EM) stimulation of the brain will be far more subtle, specific, and reliable in its production of mental effects than pills, which must make it through the stomach into the bloodstream and from there across to the brain. Yet EM patterns may be very much like drugs in that they function by simulating the body's own ionic neurochemistry, which is actually electroneurochemistry, and thus impacted, if not outright controlled, by electromagnetism.

Well over three billion years of biological evolution have produced many magnetically sensitive organisms, from bees unable to find honey if magnets are placed under their hive to migrating birds with magnetic cells in their brains and magnetotactic bacteria that orient like an animate compass by swimming toward the magnetic north pole. We ourselves have light-sensitive cells inside our bodies, a result less of design than of a legacy from free-living cells exposed to light. Light is, of course, a form of electromagnetism, and thus one could argue that responsiveness to electromagnetic fields is, along with smell that detects small concentrations of chemicals, one of the two most ancient means of perception.

Earth's magnetic field is not constant but fluctuates, in part due to interference from flares of plasma ejected by the Sun. Persinger found the number of reports of bereavement apparitions—dead people taking an encore—rose with increased geomagnetism as measured by magnetometers. Further research revealed the curious coincidence of luminous balls appearing prior to earthquakes. Persinger postulated that these balls—thought to be rare EM phenomena, somewhat like ball lightning or the aurora borealis, but produced by intense heat and pressure within the Earth's crust—were mistaken for UFOs. The "Earth lights" would change shape and color, rotate and move, and interact with moving vehicles. Indeed, Persinger, compiling statistics, found that in the six months preceding an earthquake UFO sightings, which clustered along fault lines, tended to increase. The electromagnetically sensitive brain and "geomagnetic strain" theories, taken together, suggest that electromagnetic energy luminously released along fault lines in the Earth's crust is mistaken by gullible humans for other life-forms. We are like those wasps that think they are scoring with females when really they are rubbing against orchid parts, or like that firefly I saw in the woods that tried to mate with the lit tip of my cigarette—there was a real idiot.

Since their color reflects their temperature, if they turn luminous, magnetically charged earth chunks will display different colors. Considering that driving late at night can predispose one to seeing things, such bright colors might easily qualify as an alien vastation. The luminous displays, moreover, would be expected to travel along with electrically conductive metallic cars. Although the material substrate of the luminous balls is not known, their airborne magnetism would be sufficient to disable spark plugs, stopping automobiles. Diesel engines, which use glow plugs, would not be affected.

Now to this external mix add the internal effects an Earth-burped luminous object could have on the late-night human brain. You have the ingredients for a gen-u-ine UFO sighting. When one also ponders the recentness of magnetically attractive metal cars traveling at high speeds one realizes that our ancestors could not have come up with a convincing interpretation of such events—they never had the opportunity to experience them, let alone interpret them as alien encounters. That a government that can't help leaking irrelevant sex gossip systematically covered up alien sightings now seems even less likely.

"We are empiricists," Persinger says. "We postulate that all experiences are generated by brain activity." He goes on to tell me about near-death experiences, which he has been studying, along with UFOs, for about thirty years. They are Canadian near-death experiences; they involve snowmobiles. In one, a man who worked categorizing parts in a factory was thrown from his snowmobile, suffering a concussion. He saw a very white car down at the end of a narrow road. The car had a red light behind it. His recently deceased grandmother got out and addressed him. In another, a man in a snowmobile had an accident. He was drunk. Momentarily, he lost consciousness. A figure appeared, hovering over him. "Stop drinking, you asshole," it said. The man recovered from his near-death experience but ever since he has not been able to tolerate the taste or smell of whiskey.

Real near-death experiences are not like Hollywood; they are highly individualized, not spiritual clichés. Yet all spiritual experiences seem to have a common denominator: they are electromagnetically mediated. In Persinger's reading, that common denominator belongs to this world, not the great spirit world beyond.

Chapter 12 Note
1. Persinger, 1987.

part three

eros

We humans are obsessed with sex. If sex among our ancestors had been dispensable they would have saved a lot of time. But it was indispensable. Our emotionally and physically preoccupied lineage requires sex to reproduce. However, not all life shares our pleasures. The majority of living beings reproduce without any requirement to mate. Here we try to increase our empathy with other, non-human living beings who, from the beginning, have only a single parent. These life-forms may not seem erotic but they are, in some ways, far more intimate than we.

This section is organized under the aegis of Eros, the michievous Greek god of love, sometimes portrayed with his twin, the love god Himeros (Desire), joining the goddess Aphrodite upon her birth from sea foam. Here we treat questions of the evolution of sex and gender, and the parasexual transmission of genes and genomes across would-be species borders. For sex's evolution appears to have been a rather epochal, traumatic, and, unsurprisingly, messy event. Evolutionary biologists have long puzzled over why sexually reproducing beings from maple trees to human beings would go through all the trouble to split their chromosome numbers in half, producing sperm and ova, only to put them back together again. It does seem to be the ultimate exercise in futility, frustration, and waste. Human beings have felt the paradox more directly: so much of our energy goes into the mating and dating game—wouldn't it be easier simply to divide like an amoeba? As Lord Chesterfield put it, "The pleasure is momentary, the position ridiculous, and the expense damnable." Although evolutionary biologists have speculated that recombined genes provide a great advantage to organisms navigating rapidly changing environments, protecting them also from quickly mutating disease organisms, there is a fatal flaw in this theory: clones, near-identical genetic copies, show surprising variation.

The real "reason" sex may have evolved is that, like Zeus's father, Cronos, our multimillion-year-old microbial ancestors fed on their own. Sex, at first, seems to have been an emergency response among threatened beings. Like the downed crew of an airplane crashed on a remote mountain top, they were cornered into cannibalism. Our cellular ancestors swallowed their fellows whole, avoided starvation and, in the transgressive bargain, acquired from their victims an extra set of genes. Without immune systems, some survived and grew in their merged state. Abortive cannibals with two sets of chromosomes, their own and those acquired from ingestion of their victim sibling, must have been quite common in the long past. Abortive cannibalism was common enough to occur serendipitously in modern microbiological laboratories. Genetic doubling through eating followed by incomplete digestion must of course have been far more common in the natural laboratory of Earth's ancient surface. With minor adjustments in reproductive timing, such doubled cell beings grew into life-forms visible to the naked human eye—such as, eventually, other naked humans. But they always had to, as was and is their wont, return to their primordial single-celled state. We still return to single sets of chromosomes, our single-celled state of sperm and egg cells. Our penchant for sex, without which our particular form of chemically cycling material organization would not persist, seems to be a sort of frozen accident that must be repeated every generation for us even to continue to exist.

In retrospect the hypotheses for "sex's adaptive value" miss the point that our kind of animal life-form has been sexual since its evolutionary inception. Our sexuality does not derive directly from only benefit, in other words, because most, if not all, sexually reproducing animals have no exit, no way to opt out of the ancient cycles of meeting, mating, and cell growth to make our bodies. We must return to the primordial single cell form: egg or sperm. We repeat the cellular infrastructure and life cycles of our ancient ancestors. In pursuing the beings we know and sometimes love, Cupid's arrows are dipped in the potent potion of what has worked to maintain our material organization over evolutionary time.

The Riddle of Sex

DORION SAGAN AND LYNN MARGULIS

Eden may be the preferred origin myth, but animal sexuality in truth is an outgrowth of the sex lives of protoctists (chapter 4). Multiple lines of evidence suggest that nothing was more important to the origin of sex than abortive cannibalism in our single-celled ancestors. Stress and death have accompanied animal sex since its ancient inception, at least seven hundred million years ago, in the sulfurous muds of the Proterozoic eon.

Why are so many organisms sexual? What keeps them competitive with organisms that accomplish the same end through budding or fission? Most biologists explain the existence of traits in organisms on the basis of survival value to the individual or the species.[1] Yet the story of the evolution of reproductive patterns is not clear-cut. Our ongoing efforts to understand this process illustrate a key area of research in evolutionary biology today.

At first and perhaps second glance, sex seems a superfluous and unnecessary evolutionary bother. To put it in the economic language in which biologists have described evolutionary science from its inception, the "cost" of sex—finding mates, producing special sex cells with half the usual number of chromosomes, and investing time in these activities—seems all out of proportion to any possible advantage.

Biologists have thought that sex remains because of the increased variety of zygotes that results from two parents. Sex increases variation, it was reasoned, and this allows sexual organisms to "adapt faster to changing environments than do asexually reproducing organisms." Yet there is little evidence that this is true.[2]

When the idea was tested by comparing animals that can reproduce either asexually or sexually, such as rotifers and uniparentally reproducing lizards, scientists found that as the environment varied, the asexual forms were as common as or even more common than their sexual counterparts.

Biologists need a new perspective on this important problem. We believe that sexuality in animals is a product of a history in which sex became entangled with reproduction. Sexual animals have been successful for reasons not directly related to biparental sex. Thus, we think that it is not sensible to ask, What selection pressure maintains sex in an organism? Once animals and some other organisms became committed to a link between sexuality and reproduction, in most cases there was no turning back.

How did these organisms become sexual in the first place? In discussing a topic so enriched by the imagination, we must define our terms. Sex in the biological sense has nothing to do with copulation; neither is it intrinsically related to reproduction or gender. Sex is a genetic mixing in organisms that operates at a variety of levels; it occurs in some organisms at more than one level simultaneously.

We, then, are defining sex as a union of genetic material to produce an individual from more than a single parent. The smallest known framework for sex, defined in this way, is the entry of nucleic acid into a cell. In bacterial sex, bacteria regularly exchange genes, for example, by passing genetic information as plasmids or viruses. In protoctists, plants, fungi, and animals, sex takes the form of a fusion of cells such that two nuclei from different parents join within a common cytoplasm. Figure 13.1 shows two *Stentor coeruleus* of complementary genders, in sexual embrace.

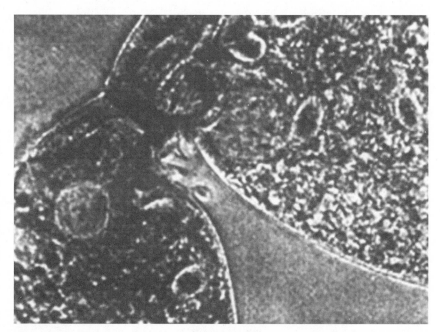

Figure 13.1 Sex in ciliates (genus *Stentor*) of complementary genders: the mates look the same to us. The death of both individuals follows the process within four days.

BACTERIAL SEX

Bacterial sex may have begun over three billion years ago, when our atmosphere lacked free oxygen.[3] Without oxygen, no ozone layer existed in the thin atmosphere to protect genetic material from ultraviolet radiation.[4] Data gathered by *Explorer 10* from Sun-like stars suggest that the output of light energy at that time may have been so great that it was a wonder that genetic information survived at all.[5]

Yet life did develop, under the pressure of constant bombardment from both benign visible and dangerous ultraviolet rays. Bacterial and viral sex must have soon followed, as a means of guarding and spreading needed genes throughout the threatened biosphere. Any organism that could not protect its scant genetic hoard in that age soon perished.

Early sex seems to have developed from a genetic repair system that could restore damaged DNA. The first repair may have happened by chance—a case of chemical desperation. Those cells that detected damaged DNA and excised it survived. As time went on such methods were refined. In photosynthetic organisms in particular, for which the radiation was both essential and lethal, repair became a way of life.

In standard DNA repair, an organism copies an intact strand to produce a healthy double-stranded molecule. This splitting and splicing is closely related to sex—the mechanism that allows the cell to accept DNA from a foreign source. Thus methods that first allowed survival in a radiated world evolved into sexual mechanisms.

Bacterial sex promoted both diversity and survival. New varieties arose as patterns for new proteins were shared and copied. Even today, toxins and ultraviolet light can revive eons-old solutions. Some bacteria respond dramatically to ultraviolet DNA damage. They immediately stop growing, release viruses or plasmids (if they have been harboring these small genetic entities), and make error-ridden copies of their damaged cell DNA (the "SOS response") so that at least some descendants will survive.[6]

Conversely, when bacteria lose the ability to deal with ultraviolet light, they often also lose their genetic recombination system. The "rec minus" mutant of *Escherichia coli* can no longer recombine; it is also hundreds of times more sensitive to death by ultraviolet radiation than its sexual relatives. The two processes, protection from ultraviolet radiation and genetic recombination, must be very closely related.

Thus the repair of ultraviolet light damage may have preadapted bacteria to sex. By rupturing genes, this energetic form of light put selection pressures on bacteria for the development of repair systems, some of which involved "adopting" DNA from neighboring cells. By the time the atmosphere developed a protective layer of ozone, splice-and-repair mechanisms had been integrated into the life of bacteria.

The genetic recombination that so fascinates genetic engineers today evolved first as a technique for DNA repair and then into the closely related sexual mechanism. Research on fertility factors, episomes, plasmids, infection, and conjugation all involve the recombining of genes; they are all forms of bacterial sex.[6]

MEIOTIC SEX

The sexuality of familiar plants and animals—the sex that is hitched to reproduction—is not the genetic splicing of bacterial sex. Meiotic sex, found in eukaryotic organisms, is an entirely different procedure that evolved after bacterial (i.e., prokaryotic) sex. Meiotic sex involves two reciprocal processes: the reduction by half of the number of chromosomes to make sperm, eggs, or spores and the fertilization that reestablishes the original chromosomal number.

While bacteria have sex under certain conditions, they never need it to reproduce (figure 13.2). Most animals and many plants, however, must undergo this complex process for the species to survive. How did meiotic sex evolve? Its origin seems tied not only to mitosis but also to symbiosis, the history of which is a fascinating evolutionary puzzle in its own right.[7] (The symbiotic origin of mitotic cell division as precursor to meiosis is just too complex to detail here; see chapters 3 and 4.)[8] Let's just say that the ultimate effect of mitosis is the distribution of genetic information in DNA-protein packages, called chromosomes. This meticulous genetic delivery system handles hundreds of times more genetic information than bacterial cells. Its efficiency institutionalized it as the standard mechanism of cell division in plants and animals.

Meiosis as a form of cell division follows a pattern very similar to mitosis; it differs in that when a cell divides, chromosomal DNA does not replicate, and the kinetochores (structures attaching the chromosomes to the spindle) are delayed in their reproduction. The result is the formation of haploid cells, destined to meet and restore the diploid number in offspring. Because meiosis never occurs in organisms that do not regularly undergo mitosis, and meiosis is a variation on mitosis, meiosis is assumed to have evolved by modification of mitosis.

The two distinct phases of meiotic sex—chromosome reduction and

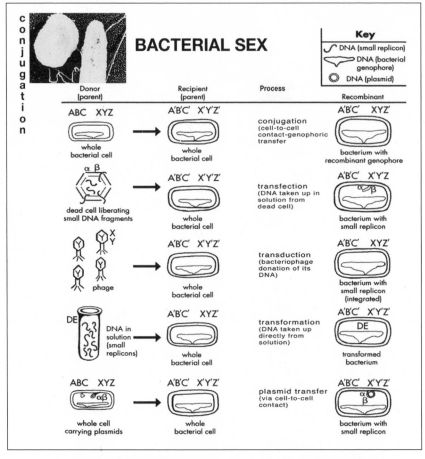

Figure 13.2 One parent (donor) gives genes to the other (recipient) with no reproduction in these several modes of bacterial sex.

fertilization (cell and nuclear fusion)—arose separately. They are still separate and occur only sporadically and irregularly in some eukaryotic microbes. However, in certain lineages, such as those ancestral to animals, meiotic chromosome reduction and the precise fusion known as fertilization became coupled (figure 13.4). Chromosome reduction probably began as a delay in the timing of reproduction of both kinetochores. By waiting too long to reproduce in mitosis, the kinetochores pulled two chromosomes to one of the two offspring cells instead of the normal one chromosome. The other cell, now without chromosomes, dies. Thus, in the very next cell division each offspring is left with only one, rather than the original two chromosomes. This resulted in a reduction in chromo-

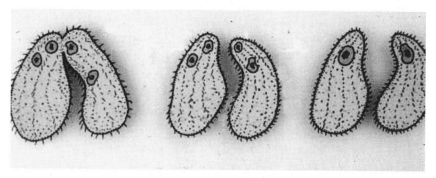

Figure 13.3 *Paramecia* mating; sex without reproduction.

some number in the offspring cells—the fortuitous evolution of meiosis from mitosis. If we regard these reproducing kinetochores as remnant spirochetes living in the chimera of a modern nucleated cell, this explanation of their duplication delay as the origin of meiosis seems likely (Margulis and Sagan, 1991).

The first cell fusion, a precursor to fertilization, could have resulted from cannibalism, where one already-mitotic microbe ate another without digesting it (figure 13.4). Microbes have no immune defense against such an internal grafting. This cannibalism would have led to diploidy, the doubled state of chromosomes that is "relieved" by meiosis. Regardless, meiosis and fertilization had to have become interlocked in a feedback cycle for today's patterns to have evolved.

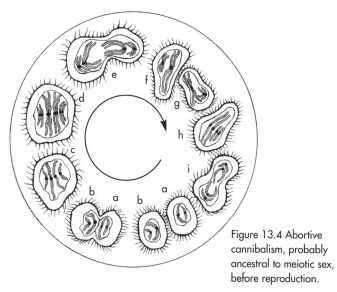

Figure 13.4 Abortive cannibalism, probably ancestral to meiotic sex, before reproduction.

Once meiotic sex became established it flourished even in single cells (figure 13.5). But why? Biologists must be careful not to jump to conclusions. The evidence shows that meiotically sexual organisms are not automatically more varied or better adapted to changing environments than those lacking meiotic sex.

When paramecia are ready to conjugate, a complex process of meiosis and mitosis produces eight haploid micronuclei, all but one of which die (figure 13.3). In the center pair, the two nuclei have focused in the right-hand pair in the ultimate act of incest. This micronucleus divides mitotically to produce two micronuclei. If a partner is available, one micronucleus will be exchanged and the pair will fuse. If no partner is available, the cell's own two nuclei will fuse. The cell achieves rejuvenation in either case.

Meiotic sex may have originated from a form of cannibalism; consider the hypothetical cycle in figure 13.4. Clockwise, two starving protists (a, b) fuse to form a double organism (c). The chromosomes replicate (d) and the organism divides (e), but the "tardy" kinetochores lag behind (d, e, f, g). The kinetochores finally replicate (h) and the cell divides again (i) to form haploid organisms (a, b).

Many theories of sex are clearly fallacious. A recurrent but dubious interpretation describes sex as some sort of genetic rejuvenating mechanism. This theory is based on the observation that among protists that reproduce both asexually and sexually, such as paramecia, asexually produced strains survive for only months, while sexually conjugating strains survive indefinitely. Yet there is a counterexample to the theory.

The words *male* and *female* have different meanings in the protist world. In mating *Trichonympha*, the organism that enters its partner from behind is arbitrarily defined as male (the lower right-hand organism, figure 13.5). In others like *Stylonychia* (not shown) or *Stentor* (figure 13.1), the mating ciliates are identical, and male-female designations are meaningless.

A paramecium that gets ready for sexual conjugation but finds no partner undergoes a process known as autogamy, in which the nucleus of a single cell undergoes meiosis and the products of meiosis from the same cell fuse in the absence of any sexual partner. This totally inbred cell line survives just as long as do its conjugating relatives that undergo two-parent sex (figure 13.3).[9]

Paramecium aurelia, for example, has one large macronucleus and two small micronuclei. The enormous macronucleus, with thousands of gene copies, usually does all the work, making messenger RNA, while the diploid micronuclei do nothing. But in preparation for the sexual event

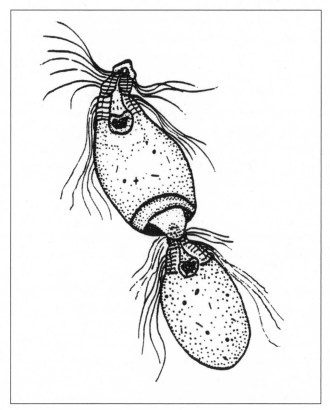

Figure 13.5 Mating *Trichonympha*. Drawing by Dorion Sagan.

each diploid micronucleus divides twice meiotically, forming four haploid micronuclei. These four divide mitotically once, creating eight haploid micronuclei. Then, in nature's typically absurd style, all but one of the haploid micronuclei die. This last one divides mitotically to create two micronuclei with exactly the same genes. If a willing sexual partner is present, conjugation occurs and one of the two micronuclei is sent to the partner as another one is received from the partner. But—and this is a big *but*—if no partner is around, the two haploid nuclei fuse. No new genes have entered the paramecium. Indeed, this self-fertilization (autogamy) renders the organism entirely homozygous. Yet the paramecium is recharged, rejuvenated, able to reproduce again for generations.

Thus it is not the receiving of genes from two parents but the meiosis that often accompanies the gene exchange that is important for survival in *Paramecium aurelia*. We think that meiosis became tied to two-parent sex and that meiosis as a cell process, rather than two-parent sex, was a prerequisite for evolution of many aspects of animals.

Meiotic sex and tissue-level multicellularity both evolved well before 541 million years ago at the base of the Cambrian Period. Meiosis seems intimately connected with complex cell and tissue differentiation. After all, animals and plants return every generation to a single nucleated cell. We believe that meiosis, especially the chromosomal DNA-alignment process in prophase, is sort of like a roll call, ensuring that sets of genes, including mitochondrial and plastid genes, are in order before the multicellular unfolding that is the development of the embryo. Meiosis has been maintained because it is connected to physiological necessities such as tissue development in the hosts. When the complex meiotic organisms survived, meiosis was taken along for the ride.

Putting these ideas of sexual origins together, our hypothesis is quite different from the accepted wisdom about the role of sex in evolution. Bacterial sex, a modified DNA repair mechanism, allows organisms to accept new genetic components as easily as one can catch a cold. Without it the more complex cells of animals and plants could not have evolved. Although the roll-call processes of meiosis are crucial, the two-parent aspect of meiotic sex is an evolutionary legacy. Animals' complex tissues and organs, not their sexuality, are the traits that were selected for. Some animals did, in certain cases, forgo two-parent sex, but they never gave up meiosis.

Ongoing work in this area opens up intriguing ideas, including the possibility of human reproduction that would circumvent the biparental sexual cycle by cloning an egg to make a person. If we are correct, biparental sex, but not meiosis, will be bypassed in such cloned people. Like many areas of biological research, conventional wisdom concerning sexual origins should periodically be reexamined.

Chapter 13 Notes

1. Futuyma, 1985.
2. Bell, 1982.
3. Cloud, 1983.
4. Margulis et al, 1976.
5. Canuto et al, 1982.
6. Sonea and Mathieu, 2002.
7. Margulis, 1993.
8. Sagan and Margulis, 1993.
9. Margulis and Sagan, 1991.

An Evolutionary Striptease

DORION SAGAN

To peel off the layers of sexual history is to attempt to pierce the mystery of sex at the center of human reproductive being. In the present archaeology of sexuality, each layer that unfolds brings us closer to the amoral innocence of sex's beginnings, exposing beneath the present the wildness of our animal and pre-animal ancestors.

I would love to kiss you
The price of kissing is your life
Now my loving is running toward my life shouting
What a bargain, let's buy it
—RUMI, ODE 3881 (BARKS, 1987)

ONE

Sex has many origins: evolutionary, sociolinguistic, and perhaps even unconscious or metaphysical origins that are not really origins at all, since they stand, at least psychologically, outside of time. Metaphysically, the conjunction of two individuals in an act of mating recalls the original split of each individual in his or her essential solitude from the universe of which he or she is a part. Thus, biology aside, the union of opposites resembles a sort of awkward "healing" of the primordial condition in which each of us finds ourselves: separate and alone.

To peel off the layers of sexual history is to attempt to pierce the mystery of sex at the center of human reproductive being. In our whimsical archaeology of sexuality, each layer that unfolds brings us closer to sex's beginnings, exposing beneath the present we take for granted the erotic pasts of our animal and pre-animal ancestors. This unraveling, this shedding of layer after sexual layer in an attempt to gaze back into the evolutionary past, itself proceeds titillatingly, in the realm of signs and images. The evolutionary striptease—a flagrant act of scientific exhibitionism—is therefore seductive with regard to knowledge. On the one hand, the sex lives of our ancestors—hairy apes, mesmerized reptiles, and cannibalizing cells—can be tentatively deduced from the circumstantial evidence of comparative anthropology, primatology, herpetology, genetics, paleobiology, and so on. On the other hand, this philosophic representation itself obeys an erotic logic, keeping separate the desire to know from the pleasure of knowledge as presence. For times past, as the spatial metaphor for time in the Navajo language reminds us, are more visible than times to come: history is not at our backs but in front of us, spread out, however mistily, for our mind's eye to examine and reevaluate. While the future remains opaque, invisible, we can, with evolutionary hindsight—with retrodiction rather than prediction—see where we have been. The translucent past can therefore be laid out ahead of us in the figure of an exotic time dancer as she begins to strip; although arguably more real than any science-fiction guesswork about the future of human sexuality, this mirage obviously is just one possible interpretation.

And now let the spotlight fall, the curtains part, and the show begin. As we watch, the evolutionary stripper begins shedding layers to reveal our sexual reproductive past, through the gray mist, to the rhythm of silent music. She looks like a fashion model, tall and thin with long thighs and makeup on her face. As we watch, her clothes come off and she stands naked, a human female of our species. But the exotic dance begins where most striptease leaves off, and she swirls before us into the past. The slender body dissolves, and a plump Paleolithic woman emerges, wearing clothes made of grass and cosmetics of clay. And then the Paleolithic woman fades into a small estrous ape-woman, with receding forehead and thin hips. The ape's pubic region is swollen and her brown buttocks flash rouge and purple. Now she turns again, shrinking into a still hairier, more unfamiliar primate. But why look? Our interest is in the tantalizing movement of exposing, rather than exposure; so let us turn away from this as yet unrated scene.

As is the case with a jealous lover rummaging through a mate's belongings for evidence of another lover, the seeker of ancestral sex clues seldom meets with straightforward success. And like the vivid scenes conjured up in the mind of a jealous lover, scenarios of ancestral sex lives—the sex lives of our forebears—also rely upon circumstantial evidence.

But if we cannot storm in and catch our ancestors evolving *in flagrante delicto*, we can, as the Motown song says, "hear it through the grapevine"—study the bodies and behaviors of live organisms for clues to our animal and microbial past.

TWO

As the stripper peels off the uppermost layer, we see that the human body itself attests to promiscuity in ape-people.

Since reproduction in many of the species ancestral to humanity was sexual, the shape of the present-day human body conforms to the sexual predilections of our ancestors: men and women's erotic choices and behaviors have helped physically shape the human body. Darwin recognized that "sexual selection" was a process as potent as natural selection but was caused by adaptation to members of the same (intrasexual selection) or opposite (intersexual selection) sex rather than to the environment at large. Human males compete with each other for access to females by having heavy testicles capable of producing many sperm—far more, for example, than gorillas. Secondary sexual characteristics such as women's breasts (apes have nipples but not true breasts), the absence in women of a distinct period of estrus or heat (as chimpanzees have), and the relatively big penises of men (compared with those of the great apes—gorillas are only about one inch long erect) have all been explained, albeit after the Victorian-era Darwin, in terms of the competitive advantage of male equipment during periods of evolutionary time when females have sexual relations with multiple partners. Over evolutionary time our bodies have changed as dramatically as clothing fashions, partly because of the environment, but also because of the needs and tastes of the opposite sex, as well as competition among each sex for select mates. Ancestral males and females could have lived together in beautiful harmony or violent strife—and no doubt did both. As Bette Davis cackled when asked what she thought about marriage, "With separate bedrooms and separate bathrooms, I give them a *fighting chance*." Indeed, in terms of the perpetuation of the sexes, connubial bliss is irrelevant: as long as sufficient numbers of males and females are attracted to each other long enough to mate, and to their offspring sufficiently to raise them, then the ancient mating and dating games will continue, even at the cost of their players.

Darwin stressed the choice of females in sleeping with "the least distasteful males" and the charm and fighting ability of males in competing for a perpetually scarce supply of females. Horns, antlers, tusks, ornate dinosaur protrusions, and even fencing and cowboy accoutrements such as swords and six-guns can be interpreted in terms of an epochal battle among males to dominate each other and win females. Yet due to the (pun intended) seminal work of University of Liverpool biologist Geoff Parker,

Figure 14.1 Whale penis.

theoretical biologists have suggested that almost as important as the competition among male bodies for possession of females before mating is the competition among sperm for eggs after mating. (Parker introduced the term "sperm competition" in 1970.)[1] And sperm competition applies to our ancestors probably more than it does to us today. Other things being equal, if two or more men copulate with the same woman within a period of about a week, an advantage in begetting offspring will accrue to the one who ejaculates the most sperm. So-called sperm competition arises because active sperm from two or more males may be found at the same time within the same female. This sets up the conditions for a contest inside the female even after she has copulated. If one male does not totally monopolize her body, or if she chooses not to mate with only one male, then a battle at the level of sperm—and their genital deployers—comes into play.

And, despite the possible sexist connotations, it is mainly sperm rather than eggs that compete. In humans, for instance, the number of sperm cells released in a single ejaculation is some 175,000 times more than the number of eggs a woman will produce in her entire lifetime. (Although this number is decreasing as environmental toxins, such as estrogenlike compounds, impair men's ability to produce sperm at former rates.) With this disparity, even marginal female promiscuity promotes conditions for a sperm race in which many enter but only a few win. You, for example, are testimony not only to a long line of improbable male suitors, but also to an even longer, and perhaps luckier, line of their more numerous sperm.

Factors favoring fertilization of one male's sperm over his competitors' likely include such things as position during sexual intercourse, number and swimming speed of ejaculated sperm, and proximity of the spermatic means of delivery—the penis—to the egg at the time of ejaculation. Copious sperm production (gauged by testicle weight), deep penetration, and an elongated penis all presumably advantage males engaged in sperm competition.

The struggle to fertilize the egg is an example of Darwin's intrasexual competition among members of one sex for access to the other. For Darwin intrasexual selection was primarily male-to-male competition. In *The Descent of Man and Selection in Relation to Sex* he referred to it as "the power to conquer other males in battle." Whereas *inter*sexual selection—"the power to charm the females"—can result in the evolution of bright colors and ornamentation (as seen in many birds, the males of whom are more brightly colored) and even, some suggest, in the ancestral human mind (with its possibly environmentally superfluous ability to make sentences and songs that impress females looking for signs of their potential mates' survival prowess), intrasexual selection produces huge bodies, sharp canines, and other natural weapons. But, given active female sex lives, *intra*sexual selection can also lead to "peaceful" adaptations, such as a large penis and heavy testicles. Darwin may have been too polite to write or too Victorian to think about this fascinating possibility.

Comparing the genitalia of men with those of the great apes, one is led to speculate that sperm competition probably played a greater role in the human past than it does today. Nonetheless, it still occurs. Sperm-competition theorist Robert Smith reports of a German woman who bore twins, one of whom was biracial—the offspring of an American GI—and the other of whom was white—the child of a German businessman.[2] This is clear evidence of sperm competition, though in this case the competition ended in a tie, because the woman produced two eggs, and dizygotic (fraternal) twins were the result. Promiscuity, communal sex, prostitution, infidelity—human sexual behavior ranging from dating to rape—set into motion the contest among multitudes of sperm from different males for scarce ova.

The alternative to entering the sperm competition is to bar other males from the contest altogether by harassing, bullying, or killing them. (For less brutal souls, marriage and elopement may also be options.) These two main strategies—sperm competition and sperm-competition *avoidance*—seem to be reflected in the bodies of our closest living primate relatives: the chimpanzee, the orangutan, and the gorilla. Chimpanzees, who can be very promiscuous indeed, produce more sperm per ejaculate and have heavier testicles for their body weight than humans. "Chimpanzees are reproductively extraordinary among the great apes. They are not strikingly dimorphic for size as are the other two species, but male chimpanzees have enormous scrotal testes, proportionately about 5 and 10 times larger than *Pongo* and *Gorilla*, respectively, and a specialized penis more than twice as long as that on the much larger gorilla. The testes of an average chimpanzee can sustain sperm production at a level that will produce at least four full-strength ejaculates/day, each containing several times the number of sperm in an average gorilla or orangutan ejaculate."[2] But the big scary gorillas and orangutans have puny penises, tiny testicles, and relatively minor ejaculate volumes. The average gorilla penis measures only three centimeters (barely over an inch) when erect; the average orangutan hard-on is only four centimeters. But that's all it takes. Think of the film *King Kong*, where the giant airplane-swatting ape climbs the Empire State Building with a beautiful screaming girl in the palm of his hand, and you will understand instantly why the bigger and more ferocious the male, the less natural selection equips him for the sperm competition.

Whereas gorilla and orangutan males are far larger than their mates, chimp males (and, to a lesser extent, human males) are much closer in size to the opposite sex. The difference in body size ("sexual dimorphism") provides more circumstantial evidence that our ancestors and those of chimps were more promiscuous than the ancestors of the gorillas and orangs. Today gorillas are organized in social hierarchies: dominant male silverbacks, named for their mature silver coats, typically control a "harem" of females, mitigating against competitive ejaculation contests by shifting their considerable weight around. The expressive orange-haired orangutan, a loner, roams through the forests of Borneo. Couples are often isolated from other orangutans. If orangutan ancestors were similarly isolated, large-penised, heavy-testicled offspring would not have reproduced with greater success than their brethren with smaller genitalia, as the opportunities for female orangs to mate with more than a single male (initiating the sperm competition) would have been very limited.

Men have relatively large penises and heavy testicles, indicating that these may have been valuable survival traits in the past. Different prehuman species suggest different levels of sperm competition. Judging from its teeth, *Australopithecus robustus* was certainly a vegetarian. Like gorillas, the vegetarian australopithecines showed a bigger difference in body size between genders than men and women do today, suggesting that they were sperm-competition avoiders. Although no one, of course, knows for sure, Robert Smith speculates that physically imposing males bossed relatively sexually faithful females in australopithecine harems. These ancestors would have been very sexist, but the males, violently intolerant of promiscuity, would not have developed large genitals. This sultanlike breeding behavior could well have undergone radical change with the evolution of *Homo habilis*, "handy man": Smith postulates that subordinate *habilis* males, scavenging meat and offering pieces of it in exchange for sex, upset the earlier breeding system. The cooperative hunting groups that began with *Homo erectus*—our most recent evolutionary predecessor—ushered in relatively high levels of sperm competition. *Homo erectus* males were not much larger than *Homo erectus* females. *Homo erectus* was a communal species that not only gathered edible plants but also hunted mammoths and used fire. Eating and sleeping together in groups—the sort of cooperative groups needed to hunt—may have made them far more social, more talkative, and better barterers than their sexually dimorphic australopithecine ancestors. And more promiscuous. It was with *Homo erectus*, Smith suggests, that people developed their relatively large male genitals.

Could male preoccupation with penis size relate to the importance of sperm competition in the human past? Pornography is notorious for its selection of males with large penises. It is difficult to imagine gorillas or orangs worrying about their penile size—if at all, males might worry about the breadth of their chests. Nonetheless, according to the McCarthys,[3] two out of three men estimate their penis to be undersized. Such worries figure in feelings of sexual inadequacy, despite the well-publicized sexological finding that female pleasure depends on external stimulation of the clitoris, which is not even directly stimulated by the penis. The McCarthys attribute men's prevalent anxiety about their penis size to several factors. First, boys catch sight of their fathers' penises at an impressionable age and worry they won't catch up. Second, glances at other males in locker rooms are made end-on: the other men's penises seem larger because a man looks at his own penis from above, a perspective that makes it seem smaller because of the shift known by artists as foreshortening. Third, penises seen in a flaccid state do vary dramatically in size; erect, there is far less variation: the average human penis across populations measures five to six inches. The McCarthys also cite a generalized male reluctance to discuss personal sexual issues such as penis size: men are more likely to tell women about their sexual worries than they are to tell other men.

It may be therefore that ancestral human beings behaved sexually more like chimpanzees than they did like gorillas or orangs. Chimpanzees in heat, like many other female primates, become pink and swollen around the genitals and rear; they undergo estrus. Females in this state are very sexually active. According to Jane Goodall, Flo, a chimp mother of four, would during estrus lift up her pink buttocks flirtatiously and copulate with virtually any male except her own sons. She often enjoyed quite a few males in rapid succession. And while human culture, language, and technology dramatically influence our behavior, polypeptide sequencing—the study of detail in proteins—reveals a closer kinship of humanity to chimpanzees than to any other living species on Earth. It may be that both we and chimpanzees derive from a single promiscuous ape ancestor.

What may have happened is that societies featuring marriage outcompeted more promiscuous social units. More cohesive internally, monogamous tribes probably were better equipped to wage violence against their less possessive, more free-loving neighbors. When sperm competition declines, violence and possessiveness become more important. In phenomena ranging from jealous rage to organized antiabortion protests we see males exerting or attempting to exert reproductive control over female bodies; such control effectively replaces the big genitalia of sperm competition as a means of ensuring male reproductive success. Losing estrus, with helpless infants to care for, the smartest ape-women would have saved themselves for provider-fathers more adept at sperm-competition avoidance (that is, violence) than at sperm competition. The importance of sperm competition therefore would have declined. (Ape-men became humans so recently, however, that the sperm-competition equipment has had little opportunity to disappear.) In this rough sketch, sexual "morality" may have appeared as a cultural phenomenon legislating the behavior of individuals moving away from sperm competition and promiscuity: good for the group, if not the individual, a way of strengthening the tribe that was once more promiscuous and thus, everything considered, less organized, especially in directing energy to destroy other troops or tribes. Morality strengthens *a society's* potential for organized violence. As Charles Darwin (1871) wrote:

> It must not be forgotten that . . . a high standard of morality gives but a slight or no advantage to each individual man and his children over other men of the same tribe. . . . [But a tribe whose members] were always ready to aid one another, and to sacrifice themselves for the common good, would be victorious over most other tribes; and this would be natural selection. At all times throughout the world tribes have supplanted other tribes; and as morality is one important element in their success, the standard of morality [will rise by natural selection].[4]

Despite neo-Darwinian protestations to the contrary, there can be no doubt that Darwin was right in his implication that not only individuals but groups of individuals evolve, and that as groups they are subjected to natural selection. Animal bodies themselves are, after all, groups of integrated cells, just as societies are groups of interdependent people. Except for disease growth such as tumors, cells reproduce "morally"—that is, under a tightly regimented discipline of physiological control. Despite the evidence of active sperm competition in the past, individuals in human society today sexually restrain themselves because of sociocultural norms stretching from widespread incest taboos to more parochial restrictions such as the celibacy vows taken by the Roman Catholic clergy. Not only in human history, but in the appearance of individuality in evolution generally, restricting reproduction of individuals may be said to give an advantage to the reproduction, or spread, of the societies to which those individuals belong.

In the hundreds of thousands of years preceding history, people were hunter-gatherers. Evolutionarily speaking, our sexual psyches are probably still responding to life in those ancient times. In this long period prior to civilization, there were far fewer people on the Earth. Male anatomy today attests to more promiscuity in the past; large penises and testicles presumably arose during times of more intense sperm competition, perhaps in communal fire-using humans of the aptly named species *Homo erectus*, or maybe even prior to the hunting-gathering period. In the form of Lucy, the fossil record documents a prehuman female pelvis indicative of an upright walking posture. Perhaps already by four million years ago, as upright, four-foot-tall, chimp-faced australopithecines, human ancestors had evolved large genitalia in the context of sperm competition. And because primate estrus is so widespread, ancestral women also probably went into heat, swelling up pubically and changing colors, thereby attracting numerous male suitors. Not only loss of estrus but the appearance of female breasts represent an enigmatic change in the prehistory of femininity. For, at first glance, breasts (whose subcutaneous fat has no direct relationship to milk supply) and the loss of estrus both would repel males, turn them off: to ape-men both such traits would be associated with pregnancy and hence a temporary lack of fertility. The loss of estrus and the development of breasts would have hidden ovulation—and the oval prize of the sperm competition. But why would the woman—or her body—want to hide her fertility? Evolutionarily, the "reasons" ape-women hid ovulation by concealing estrus could have been similar to the reasons a married woman does not dress provocatively: she avoids her husband's jealousy and unwanted male advances. So, too, primordial women who concealed estrus would have benefited by escaping unwanted sexual attention while increasing the chances of commitment from a single, able man. All humans alive on Earth today may come from females who concealed estrus and ape-men who, though adapted to sperm competition, gave it up, at least partially, in order to obtain sexual favors from ape-women. And the ape-women themselves were probably competing—though not for sperm, which was always in abundant supply, but for help with their crying infants.

THREE

Voyeurism resumes, and we catch the stripper in another phase of our out-landish act. Now a primate with a vaguely chimplike face, clever eyes, and claws, she elegantly casts off her current appearance to reveal an impish-looking reptile about the size of a dog and with an intelligent, if not wholly affectionate, expression.

Peeling off the second layer, we expose, beneath the sexually selected body, the "reptilian" brain—an ancient part of our anatomy that we share not only with the apes but with all mammals and reptiles. The R-complex, as it is called, seems conserved—an instinctive, close-to-the-genes control center still infiltrating our rational consciousness, dragging down the angelic to the level of the human, subverting humans into beasts.

If anatomy is destiny, as Freud noted, then this phase of the evolutionary striptease is still more tantalizing than what has gone before.

Here, the evolutionary stripper sheds the clothes that are her ape and mammal bodies to reveal a brutal, cold-blooded, and calculating reptilian psyche. The reptile brain appears to be fixated on sex and violence and to rule the ritual and agonistic behavior—"aggression and submission, territoriality, hierarchies, display, threat, fighting, and vocalizations"—of modern reptiles.[5] For now one may cast much of what goes on at frat parties, rock concerts, and military parades as latter-day manifestations of the reptilian brain. The reptile "within us" appears to exist in a predominantly visual realm that is in some sense timeless—the waking state of living reptiles being perhaps similar to our nocturnal dreams. The continued emphasis on sight in the reptilian brain may come from the lack of a keen sense of hearing and smell in most reptiles. Reptiles process vision more in their retinas than in their brains; their communications and signification would be instinctive, "dumb"—more like a form of sign language or writing than protracted speech. Yet even metaphors of sign language or body language inadequately describe reptilian thought, for it is in the transition from reptile to mammal brains that the mammalian ability to perceive the passage of time probably occurred. Linearity itself—the "origin" of myth, language, and all evolutionary stories—would then begin at a certain point on the very time line that, paradoxically, it produces.

Neurobiologist Paul D. MacLean has pioneered the scientific description of the human brain as triune, divided into three sections reflecting our evolution from less brainy ancestors. Freud, starting about 1900, developed his "first topography," in which he distinguished among the unconscious, the preconscious, and the conscious. Later, about 1923, he mapped out a "second topography," which distinguished among *das Es*, *das Ich*, and *das Uberich*, usually translated as the id, the ego, and the superego, though perhaps more faithfully rendered as the "it," the "I," and the "over-I." Ironically—though he stated that "anatomy is destiny"—Freud stressed that his carefully delineated maps were not descriptions of locality but metaphors for the complex workings of the human mind. The biological thinker Jakob von Uexküll used the word *Umwelt* to refer to the cognitive world, the slice of its environment that each species characteristically internalizes.[6] But MacLean in his triune description shows that the human *Umwelt* is really three in one, the physiologically discrete but mentally overlapping worlds of the triune brain.

The most human and evolutionarily recent part of the brain in this schema is the neomammalian cerebral cortex, the external gray matter, which presumably gives us language as well as large heads. Beneath this is the paleomammalian brain, which we share with all mammals from horses to hamsters. The paleomammalian brain seems to govern characteristically "mammalian" emotions, such as melancholy and parental tenderness; MacLean even suggests it is implicated in the enlightenment of aesthetic, scientific, or mathematical discovery—feelings that, like religious awe, convince subjects of the correctness of their thoughts. The old mammalian brain is thought to mediate between the neocortex and the even more "primitive" striatal structures below it.

The striatal complex is "a basic part of the forebrain" consisting of "olfactostriatum, corpus striatum, . . . the globus pallidus, and satellite collections of grey matter," according to MacLean.[5] It is called the R-complex because of its striking resemblance to the entire forebrain of reptiles. The stain for cholinesterase, an enzyme that breaks down the neurotransmitter acetylcholine, vividly colors and delineates the R-complex not only of reptiles but also of birds and mammals; histochemistry reveals the R-complex to be a uniform brain entity. One histofluorescence technique developed in 1959 causes the R-complex to glow bright green, showing the presence in it of the neurotransmitter dopamine. The R-complex also contains an abundance of serotonin, a neurotransmitter implicated in hallucinogenic LSD experience, as well as mood generally, and of opiate receptors, which are acted upon by synthetic painkillers such as morphine and heroin. In MacLean's view the R-complex is a basic module of vertebrate brain physiology elaborated by evolution. It appears that in the R-complex we are dealing with an evolutionarily very important part of the human anatomy—perhaps even the physiological site of the unconscious mind.

Mammals and dinosaurs both evolved from an earlier group of mammal-like reptiles. Reconstructions of fossil bones reveal creatures that looked like dog-size lizards or weasel-like reptilians. Fossils of such reptiles, belonging to the class Synapsida, abound on every continent but Antarctica; one paleontologist calculates that some 800 million skeletons of mammal-like reptiles exist in the Kaarroo beds of South Africa alone. Evolving about 250 million years ago, these creatures expanded prodigiously during the Permian and Triassic geological time periods. They covered the Earth. But the fossil record suggests that when the swift and vicious thecodonts—the forerunners of the giant dinosaurs—evolved, only a few mammal-like reptiles survived. Unable to defend themselves against their increasingly brawny and ferocious cousins, the earliest mammals presumably took to the nightlife, hiding in bushes and in the dark, and escaping to cooler climes, where they were free from molestation. These remote human ancestors were physically incapable of competing with the dinosaurs. Those that survived had to evolve expanded sensory modalities—particularly the sense of hearing, which would warn them of the approach of predators and the retreat of prey in their newly nocturnal world. Nonetheless, our sexually reproducing four-legged ancestors must have continued to share with monsters such as *Tyrannosaurus rex* a certain narrow focus on brute survival, on killing, avoiding being eaten alive, and fighting in order to mate, rape, or avoid forced copulation.

Experiments reveal the workings of the R-complex in modern animals ranging from lizards to squirrel monkeys. For example, if one hemisphere of the striatal forebrain of a green *Anolis* lizard is surgically impaired, and one eye is covered while the eye connected to the injured part of the forebrain is left exposed to a rival lizard, the lizard will see his rival but make no ritual display. The connection to his R-complex has been severed; he does not respond with aggressive behavior to sexual competition. But if the other eye is covered—the unimpaired eye connected to the intact part of the R-complex—the reptile reacts with species-typical challenge or territorial displays: he pushes up with his feet, swells out his throat fan, and changes his position so that his imposing long side, his profile, is exposed to his rival. In a word, he makes himself *big*. The normal *Anolis* lizard with intact R-complex is like a man who puffs out his chest or stands over an adversary to get a psychological advantage. He is cold, predictable. You might even call him macho.

Even to us, let alone to the simpler reptilian mind, parts can appear as a whole. (If language works by replacing parts for wholes, by synecdoche and similar figures of speech, is it not an elaboration of the displacement and condensation already at work in the reptilian unconscious, in the shifting scene of dreams?) Enlargement of a part, such as puffing out a throat fan or turning one's body to occupy a bigger slice of the rival's field of vision, may be a form of protolinguistic deception. Such acts trick potential enemies into thinking they see an animal bigger than the one actually present there. Certainly such displays are an economical way to frighten an enemy: you *pretend* to be big.

Of course, it is still more impressive actually to *be* big. In fact, the whole dinosaur drama, culminating in the extinctions of the dinosaurs some 65 million years ago, is a story of gaining advantage by increasing size. Could it be that the desire for bigness is coded into the human R-complex?

Translated into English, the R-complex may contain messages such as "Avoid animals bigger than me" and "Try to seem as big as I can." Is it possible that the human male fixation on big penises is a manifestation of a primordial commandment or desire for bigness still lurking in the reptilian brain?

But the ancestors of mammals benefited precisely from their smallness, as it forestalled the effectiveness of the bigger-is-better strategy. Ultimately, they regrouped and became masters of a different order of bigness: big brains.

Appearing big may even have been important in that other branch stemming from the reptiles: the birds. The evolution of birds has long been an evolutionary mystery because birds could never have evolved wings "in order" to fly; at first the wings must have been mutant limbs fortuitously valuable for something other than flight. Some evolutionists suggest that birds may have used pre-wings and mutant feathery scales as a sort of insulation, a means of temperature control. But were not wings also used by birdlike reptiles to frighten their rivals and deceive them as to their true size? One can imagine that the raising of scaly wings was an effective premilitaristic display, something like raising a flag or turning sideways. The wings-to-be could have cast fearful shadows, petrifying other animals in broad daylight. Suddenly lifting, they could have simulated the approach of much larger beasts to animals with R-complexes unable to make such distinctions. The use of such wings would—like the growling of a small animal that mimes the vocalization of a larger one—be a kind of macho bluff, a primordial scare tactic evolved, like insect mimicry, at the level of the body rather than speech. Flightless, such wings would lie.

Of course primates, thanks to our reptilian endowment, are no strangers to "ritualistic" behavior. The social behavior of squirrel monkeys—creatures far closer to us on the evolutionary tree than ancient reptiles or *Anolis* lizards—was examined in one of MacLean's seminal experiments. Although they do not speak, squirrel monkeys communicate on a preverbal ("physiological") level. MacLean calls the nonverbal vocal, bodily, and chemical signals in these monkeys "prosematic." He was particularly interested in a gesture that figures heavily in seduction, aggression, dominance, and submission: male squirrel monkeys use their erections "polysemously" to show aggression, to signal sexual desire, and as a form of platonic greeting. Members of one species invariably hold their erect penises up to their own reflection in a mirror; apparently they are trying to frighten away a rival male. Because they predictably display to their image, MacLean made a systematic study of the effects of brain ablations on their prosematic erection behavior, which he suggests is connected not to more recent parts of the brain, but to the irreducible R-complex. Bilateral lesions of the neomammalian and paleomammalian parts of the primate forebrain caused either no or only a transitory effect on the erect penis display: this kind of behavior was not related to the newer, mammalian parts of the brain. Bilateral lesions of parts of the R-complex, however, "short-circuited" the displays: R-complex-impaired squirrel monkeys no longer exhibited to the mirrors. MacLean was fascinated to find that the squirrel monkeys, apart from not holding their penises up to their images in a mirror, otherwise acted virtually normally.

These experiments demonstrate that the intact R-complex is involved in sociosexual behavior not only in reptiles but in intelligent primates. Outside a certain Edenic or serpentine symbolism, the penis is not usually considered "reptilian." Yet if we look in the evolutionary mirror, monkeys—at least MacLean's squirrel monkeys—are performing a particular ritual of sociosexual exhibitionism at the behest not of their higher mental faculties so much as of their dirty old R-complex. The presence in human beings of an R-complex very much like the generalized forebrain of reptiles, from garden lizards to crocodiles, suggests that the core of our behavior is still, in a sense, reptilian. However cultivated, orderly, and rational we act, a part of the brain stalks in the shadows like a poisonous, fork-tongued snake.

FOUR

In the third phase, the evolutionary stripper slips out of her snakeskin skirt and reveals a still more primordial level of sexuality, the ancient cannibalistic gorging, writhing fusions, and literal duplicity of ancestral cells.

Wet and slippery, sex doesn't fossilize well. Unlike trilobites on Precambrian shores, insects trapped in the precursors of amber, or ape lovers dragging their feet (and thereby leaving footprints) in a romantic stroll through the drying mud, the cellular events at the heart of sex are rarely preserved in the rock record. Indeed, the shedding of clothes by two lovers tends to bring to the surface the warm, wet, salty, sticky environment of the earth as it would have appeared to an observer living three billion years ago. Life presumably began in warm, shallow seas; the earliest communities of life were probably sticky mounds, microbial "mats" petrifying into rounded seaside stones called stromatolites. Orgasm returns the body to the primordial state of cells swimming in their marine environs, to a time when life had not yet definitively moved to land via the development and incorporation of hard, durable substances such as lignin, a major component of wood or shell or bone.

Some of the oldest unmetamorphosed rocks on Earth contain fossils of cells caught in the act of dividing. The South African fossil beds bear rocks that, cut into thin sections and viewed under the high-power light microscope, reveal traces of cell reproduction by fission, or division. But such fission is virtually the opposite of fertilization, or sexual fusion. No fossil yet found preserves the intricate details of mitosis (cell division that yields a perfect copy of the parent cell) or meiosis (cell division yielding sperm or eggs with only half the parental number of chromosomes, preparatory to fusion). It is because of this second process, meiosis, that we must seek out and join up with the opposite sex to make a new fused, or fertilized, cell if our genes are to be represented in the next generation.

Except for red blood cells (which have nuclei and then extrude them), all the cells in the human body have nuclei, or are eukaryotic. Nucleation is characteristic of those cells that—copied millions of times—make up the bodies of all plants, fungi, and animals. Protoctists are also made of nucleated cells. And the protoctists—slime molds, colonial algae, ciliates, and so on—are still undergoing the sort of cellular experimentation that led to the origins of meiotic sex in the ancestors of plants, fungi, and animals. Evolutionarily, protoctists were clearly ancestral to plants, fungi, and animals, just as bacteria, tiny cells upon which all other life is thought to be based, were ancestral to protoctists.

Since meiotic sexuality is present in some protoctists, absent in others, and represented in a sort of half-hewn or midway state in still others, it seems clear that the sexuality that extends to the human and brings man and woman together began in these unicellular and multicellular organisms, which are usually invisible to the naked eye.

The question then becomes, How did the protoctists—these microbes, more complex than bacteria but less complex than the colonies that became the first animals—ever hit upon the trick of doubling up their nuclei, chromosomes, and genes every generation? And what is the point of this doubling, when the doubled cell is only to divide in half again? Evolutionary biologists have debated such questions at great length. The potential answers are complex and technical, often revolving around the idea that meiotic sexuality must have some great benefit, for example, that it must somehow "speed up" evolution or have another function that keeps it from disappearing.

In fact, aspects of protoctist sex—meiosis and fertilization—are inseparable from animals as we know them. While all-female species of lizards and rotifers reproduce parthenogenetically—or by the "self" alone—even they remain committed to the prophase part of meiosis. Such virgin-birthing animals in effect fertilize themselves, but they too always make cells that return to the single, haploid condition each generation. So meiotic sexuality may not have a "reason"; it may simply be an indispensable part of our sexually reproducing being. But, assuming that all organisms originally cloned themselves, how did we get stuck as multicellular, sexually reproducing beings?

Observations by L. R. Cleveland (1892–1971) at Harvard University led him to reconstruct a scenario for the origin of meiotic sex, as we saw in chapter 13. Meiotic sex may have begun when ancestors of animal cells got caught in cycles of eating each other, doubling, and then dividing again. Theoretically, these protoctist ancestors were cannibals, tempted by starvation and their succulent neighbors. During the eating part of the cycle, a cell would devour a member of its own species, a conspecific, as many do today if starved. But the cannibalism would be partial; as Cleveland actually observed in contemporary protoctists called hairy mastigotes, only the nutritious inessential parts, not the genes and chromosomes of the nucleus, would have been digested. Then the nuclear membranes merged to form a single membrane, and the two-in-one cells remained alive and even were aided in certain environments by their incomplete cannibalism and doubled condition (see figure 13.4).

In the second part of the cycle, the doubled cell divides meiotically—without reproducing its genes first. The result of this "mistake" in cellular timing would, in some cases, be two "halved" cells, each now with one set of chromosomes, as in the ancestral state (but with a different combination). Such cycles, though useless in themselves except as temporary adjustment to changing conditions such as seasons and drought, would have been important in the genesis of the cell colonies that were evolving to become the first animals.

Today animals cannot dispense with the return every generation to cells with only one set of chromosomes: the sperm and egg cells that still resemble free-living protoctists, what with their undulating sperm-tail-like appendages and single set of chromosomes.

FIVE

The evolutionary stripper now discards her glittery top to reveal yet another, fourth level of sexuality, consisting of liquid patches of bargaining bacteria, primordial swingers who traded nothing so dispensable as partners; rather, they traded their identity itself: their genes.

Sex in its biological sense exists even beyond the eukaryotic level of a primordial duality for whom eating and fertilization were perhaps the same sensation. Beyond the polymorphous perversity of the undigested nucleated cell are the symbiotic adventures of bacteria. The nucleated cell itself comes from different types of bacteria symbiotically and quasi-sexually merging into new entities. In nature, bacteria attack, attach to, and penetrate each other; and, living in dense collectives, under the widest variety of conditions, they often continue to trade their genes. For example, genes for photosynthesis have been found, well outside their milieu, inside parts of cells known as mitochondria, where they serve no conceivable purpose; in this manner molecular biology attests that parts of bacteria have roamed and that genetic interchange occurs not only between but within organisms. Embryology, epigenesis, ontogeny—the whole adventure of individual growth from zygote to sexually mature adult is a kind of ecological self-organization of "moral" bacteria that takes place in and as the organism. Jumping genes, "redundant" DNA, nucleotide repair, and many other dynamic genetic processes exploit the same ancient bacteria-style sexuality that evolved long before plants or animals appeared on Earth. This sexuality arose, perhaps, from systems of DNA repair that evolved in cells damaged by solar radiation.

Bacterial sexuality fundamentally differs from the sexuality of so-called higher organisms because it occurs independently of reproduction, crosses "species" barriers, and involves, in principle, the sexual sharing of genes by bacteria all over the world. Indeed, Canadian bacteriologist Sorin Sonea and his colleague Leo Mathieu point out that bacteria, since they are able to trade genes freely across would-be species barriers, are not really divisible into (or assignable to) species at all.[7] Instead they form a global "superorganism" whose bodily contours are those of the biosphere itself. If so, this superorganism must be considered "sexual" because it is continually trading genes, although this does not lead to any offspring as we think of them in sexually reproducing species. Sonea presents the global community of bacteria almost as if it were an immortal god: Why would it need children? This masturbating superorganism has already survived for the last four billion years, producing, among other fantasies, humanity.

SIX

Finally, to a music of silence, the evolutionary stripper completely disrobes: she takes off everything.

The evolutionary stripper is a curious creature: ever changing, able to assume new forms, her soul is as blank as this paper, her fishnet stockings as simple and dense as these words. It might seem that with the genetic exchange in the most elemental living beings, bacteria, we reach the end of the evolutionary striptease. But this is not the case. Nature, said the philosopher Heraclitus in a famous fragment, loves to hide. In the beginning, and perhaps the end, was the word. The evolutionary dancer, the exotic chronicler of our striptease, discards her animal appearance to unveil our ancestors' erotic past and, beyond that, the essence of our being. But she cannot quite do it. Instead, she finally shows herself for what she is: a paper dress, clothes beneath clothes, a nudity of pure words. She is bottomless; we see nothing but what we want to see; the dance never ends.

Chapter 14 Notes
1. Parker, 1970.
2. Smith, 1984.
3. McCarthy and McCarthy, 1998.
4. Darwin, 1871.
5. Greenberg and MacLean, 1978.
6. Uexküll, 1926.
7. Sonea and Mathieu, 2002.

Vive la Différence

DORION SAGAN

> Here general principles of animal sex explained in ancestral protoctists (chapter 13) and narrated through time in the striptease (chapter 14) are applied to a single case: *Homo sapiens*. Remember, this is only one example of sexual preoccupation among some thirty million estimated species alive today!

Western thought about sex—from the story of Eve to Aristotle's belief that girl babies arise from cooler sperm—has been tainted by the notion that the female is a kind of imperfect or unfinished male. Medical science, however, has gone from treating women as though they were simply smaller men to realizing that gender confers many more differences than those that are related to reproduction. Visible gender differentiation begins at about thirteen weeks in utero; earlier, all embryos, with their webbed fingers, seem to be females.

In contrast to the feminist premise that women can do anything men can do, science is demonstrating that women can do some things better, that they have many biological and cognitive advantages over men. Then again, there are some things that women don't do as well.

One of the less visible but theoretically very important differences is the larger size of the connector in women between the two hemispheres of the brain. This means that women's hemispheres are less specialized: a stroke that damages the left side of the brain leaves a man barely capable of speech, while the same damage to a woman's brain is far less debilitating since she can use both sides for language. Although there is no hard evidence, the larger connector may also account for a woman's tendency to exhibit greater intuition (the separate brain halves are more integrated) and a man's generally stronger right-handed throwing skills (controlled by a left hemisphere without distractions).

Mary Catherine Bateson, the cultural anthropologist and a former dean of faculty at Amherst College, has described women as "peripheral visionaries," able to follow several trains of thought (or children) simul-

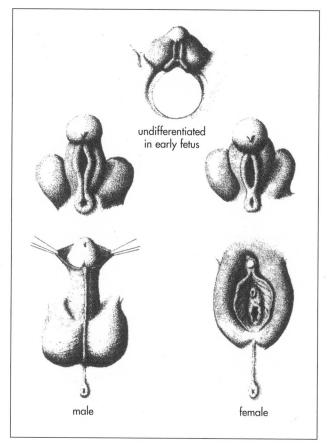

undifferentiated
in early fetus

male

female

Figure 15.1 Fetal genitalia. Drawing by Christie Lyons.

taneously. Men, by contrast, seem more capable of focusing intensely on single topics. Our strengths, then, come from our differences rather than from our similarities.

Science and medicine are finally realizing that the differences that exist between men and women necessitate developing distinct therapeutic treatments addressing the specifics of our physiology. For example, some doctors, such as Susan G. Kornstein at the Medical College of Virginia's department of psychiatry, are advocating the use of sex-specific assessment and treatment of psychiatric disorders, like depression.

Kornstein[1] points out that while depressed men seem to respond best to drugs that affect two neurotransmitter systems, those involving norepinephrine and serotonin, women respond better to drugs that affect only the serotonin system.

These differences in the therapeutic benefits of drugs not only underscore the need for medicine to go beyond giving women tapered doses of whatever is being prescribed for men (a latter-day offshoot of the women-as-incomplete-men theory) but also support the idea that men's and women's brains do not function the same way.

Indeed, it is not only our brain functions that apparently diverge, but just about every aspect of our physiology. The way we metabolize alcohol and drugs, the way our circulatory system works, and how resistant we are to infection are all affected by our sex.

Why? Hormones.

In utero, girls and boys are chromosomally different; one might wag that the determinant of maleness, the Y chromosome, named for its shape, is "missing" something that the female determinant, the X chromosome, has. But they look identical. The development of characteristic male and female sexual genitalia at birth and of secondary sexual characteristics like breasts during adolescence result from influxes of hormones, including estrogen and testosterone.

But the hormones we once thought were important only for pregnancy, lactation, and sexual drive have profound effects on just about every organ in the body. In fact, the reproductive organs, which from a biologist's perspective are our only reason for existing, control and contribute to everything from mood to how cholesterol is used in the body.

Assigning such an important role to the reproductive organs is not new to our belief system. In ancient Greece, women who were classified as having nervous or "hysterical" disorders were thought to be suffering from an upward dislocation of the womb. Treatment for nervousness and hysteria entailed, among other things, trying to repel the womb back into place by applying noxious-smelling odors to the mouth and nose.

As a few women can testify today, the perception that the reproductive organs caused hysteria later manifested itself in the widespread use of hysterectomies and ovarectomies to treat behavioral disorders among American women during the early part of the twentieth century.

Science and medicine have historically used biologically based sex differences to justify obvious acts of misogyny. It is not surprising, then, that a natural response has been for women to insist on equality implicitly based on the assumption that the sexes are essentially the same.

But women may be just as ill served by a medical profession that treats men and women as equals as by one that follows what Dr. Rudolf Virchow, a famous nineteenth-century German doctor (he was the first to describe leukemia and is regarded as the founder of cellular pathology), believed. As Dr. Virchow put it, "Woman is a pair of ovaries with a human being attached, whereas man is a human being furnished with a pair of testes."

Recent research demonstrates that while men begin to suffer from coronary artery disease earlier in life than women do, women are more likely to die of coronary complications once they are afflicted. Men are also more prone throughout most of their lives to high blood pressure, but as women get older, this advantage disappears.

The delayed onset of cardiovascular disease in women may be linked to the fact that the female hormone estrogen, which is produced mostly by the ovaries, protects the circulatory system from disease. Differences in the quantities of estrogen, which is essential for organization and maintenance of tissues and organs in both sexes, play an important role in brain development and appears to be the reason that men's brains are bigger but women's brains have more neurons.

Estrogen makes blood vessels more elastic, stimulates them to expand and allow good blood flow, and prevents cholesterol accumulation on the inside of blood vessels. As women age, however, they lose the protective benefits of estrogen because, in a rather dramatic fashion, their bodies stop producing it.

At the same time, some treatments that are used to prevent cardiovascular disorders—aspirin, for example—are less effective in women. Dr. Marianne Legato,[2] of Columbia University College of Physicians and Surgeons, notes: "Although aspirin use is associated with less frequent myocardial infarction in both men and women, it does not decrease the risk of stroke in hypertensive women, as it does in men."

There are a number of naturally produced compounds that fluctuate more in women than in men: steroids, for example, which are infamous on the street for their simultaneous role in developing muscles and shortening tempers. It turns out that steroids, a class of compounds that includes sex hormones, may play an important role in the mood swings of menstruators. These hormones directly affect brain cells. The neuroactive steroid allopregnanolone, made from progesterone, dampens the sensitivity of brain cells; it works like benzodiazepine drugs, most familiarly Valium. When the progesterone level is high, a woman is calmer. When it is low, she may feel more anxious and irritable. Moreover, women with premenstrual syndrome (PMS) become insensitive to the calming effects of Valium-like drugs.

There is a growing consensus that these steroids produced by the sex organs are responsible for the greater incidence of mood disorders and depression in women. And a growing body of research is pointing to a role for other, similar steroids in memory, stress, and alcohol abuse.

In keeping with the increasing recognition that some powerful mind-altering substances are internally produced by hormones, it is no wonder that adolescence is often a time of emotional turbulence. You cannot "Just say no" to your body's own genetically timed release of mood-altering sex hormones at puberty.

What society considers "recreational" drug use, which often begins at adolescence, may sometimes be motivated by an effort to self-medicate, changing or reversing the effects of sex hormones and neuroactive steroids. The notorious mood swings of adolescents may very likely reflect the body's adjustment to new concentrations and combinations of these compounds.

Lester Grinspoon, an associate professor of psychiatry at Harvard Medical School and an advocate of the medical use of marijuana, points out that marijuana has long been known as a palliative for the psychophysical pains of menstruation. Queen Victoria, according to her doctor J. R. Reynolds, used it for that purpose. Curiously (and since disputed), one of the few medical studies on marijuana suggests that its use lowers testosterone levels in men.

Perhaps this drug, among others, interacts with or works in a similar way to the hormonal and neuroactive steroids. In any case women, who are twice as prone as men to depression, and who have a higher body-fat-to-muscle ratio and more hormonally distinct brains, cannot be expected to respond to drugs, legal or illegal, in the same way men do.

The sexual distinction that biology traces to chromosomes and hormones also applies to culture and language. I recall, for example, being put in the girls' group at a day camp as a child because my first name was assumed to be female.

Evolutionists believe that the first sexual reproducers were unisexual cells that became involved in cycles of merging and separating. The first fertilizations probably occurred among starving microbes that cannibalized but did not completely devour each other, becoming instead two-in-one cells.

Sexual differences evolved gradually over hundreds of millions of years. With these differences came ways of recognizing them. In many species, including humans, the gametes, or sex cells, of the females became fewer, bigger, and more sedentary, while those of the males became smaller, faster moving, and more numerous. But in humans, while the female sex cells, or ova, are far larger than the male gametes, or sperm, full-grown men are bigger than full-grown women.

The cultural ramifications of body size have been considerable, including the virtual absence of rapes committed by women. They may also have influenced the development of greater female cunning and social acumen to mitigate four million years of male bullying.

In our patrilineal culture, the family name is usually that of the man. Biology tells a more matrilineal story: mitochondria, the tiny DNA-containing oxygen-using inclusions in all of our cells, come solely from our mothers. Nonetheless, culture remains, for lack of a better term, male dominated. The French psychoanalyst Jacques Lacan even argued that all speech is part of the "symbolic order": the largely negative, male realm of language and rules that supplants the original affirmative closeness of mother and child.

The psychologist, Professor Theodore Roszak of Hayward, California, has been exploring what he calls the "twisted sexual politics of modern science." He argues that science insidiously reinforces a partial male perspective. "Hard" sciences, like physics and chemistry, Roszak[3] contends, are venerated, while "softer" sciences, like anthropology and psychology, are disparaged. "Macho science," he argues, leads to bizarre fictions like selfish genes and cannibal galaxies. Female perspectives, he says, offer science new balance and openness.

From sex among equal single cells to male feminists offering cultural critiques of science's rhetoric, we have learned that the two sexes, subtly different, develop differently, respond differently to certain drugs, and see the world in different ways. As the French say, *vive la différence.*

Chapter 15 Notes
1. Kornstein and Clayton, 2002.
2. Legato, 1998.
3. Roszak, 1999.

Candidiasis and the Origin of Clowns

DORION SAGAN WITH LYNN MARGULIS

We are profoundly ignorant of the extent to which microbes influence our lives. Who would have imagined that silent, sub-visible white yeasts led to clown makeup?

PAPILLOMA, FATAL SEX, AND *CANDIDA*

Let's be serious for a moment.

HPV, the human papilloma virus, now infects 30 percent of the U.S. population. Most viruses and bacteria don't need sex to reproduce. Indeed, for the stunning blue microbe *Stentor coeruleus*, the mating act is a coupling that, without exception, results in death for both genderless partners. Why they do it is not clear. Perhaps it is atavistic, a throwback to times when they did sexually reproduce. In any case, four days after their thirty-six-hour copulation, both mates, indistinguishable from each other, shrivel and die. Reproduction remains the privilege only of those stentors that avoid sex altogether. So, too, may sex in us, the human primate, be on its way to extinction. Even though we ignore them and are literally ignorant of their influence, microbes may be stronger determinants of our behavior than are our genes. What are "microbes"? Simply organisms not seen with the unaided eye . . . or at least not understood unless viewed through a microscope.

Most microbes (fungi, bacteria, and small protists) don't need sex to reproduce. Indeed, microbiologists Sorin Sonea and Leo Mathieu[1] suggest that the natural history of bacteria is so bizarre that, had they been discovered on Mars, they would have been considered alien life-forms. The sex lives of fungi are even more astounding.[2] And yet large, familiar organisms evolved from matelessly growing bacteria. These beings do not die but divide, in principle immortally. This has brought biologists to the realization that death—intrinsic to the cell colonies we call animals—was the first sexually transmitted disease.

Virtually all autopsies of women dead from cervical cancer reveal the HPV to be up there inside their uterine cells. But the elusive HPVs (now in sixty varieties) rarely appear in men. When they do, the microscopic viruses are detected not at the tip but at the base of the penis. So, as no dunce cap can block madness, no condom can really stop this virus from spreading. Unless caught early by pap smear or prevented by vaccine, cervical cancer can lead to death in as few as three years.

I write now, free—as far as I know—from HPV but with another microbial affliction. Ragged corners of my mouth give way to fading cracks at the edges of my lips. The microbe-infested sores seem to target signs of my inner serenity: they obliterate my smile, obscuring even potential smiles right off my face. Okay, fine, they win—I won't smile. My eyes bulge. Even from a distance the cracks at the corners of my mouth make me seem menacing. I like my appearance in the mirror these days even less than usual. My problem is yeast, a yeast called *Candida albicans*. The physicians call the disease associated with this persistent little bugger with a long history "candidiasis."

I have a theory. I call it my "*Candida* theory of the clown." I suggest that the clown's sad smile is a medieval legacy: this particular fungus, *Candida* (by the way, all yeasts are fungi), attacks fools in their smiles. *Candida* loves to grow on warm, wet people when it can. It grows best where its food and water are plentiful—like near the mouth. Viruses, like the cold sore culprit, herpes, may join thriving *Candida* at the mouth's corners. Sometimes *Staphylococcus* bacteria add to the fray. Redness, cracks, and sores spread around the edges of the lips. I dreamt of hapless clowns, cliché of clichés, their lips dragged down in a curve like hopeful missiles launched skyward but returned, by gravity or the grumpiness of God. I thought of clever viruses spreading themselves by increasing the pleasure of sex, replicative geniuses as infectious as laughter. The earliest clown's affliction, I imagine, began when he was drunk each night. He tried to forget, to ignore the mean little red spots at the lips' corners. If I'm right, the cultural caricature, even in horrendous clown paintings, cliché of clichés, was authored in the beginning by strategic joint microbial action: *Candida* mostly but also sometimes herpes, *Staphyloccocus*, and other miscreants. By the time the mouth sore is large enough to be seen as a reddened mouth corner, the population of *Candida* yeast at the mouth's edge is huge.

These insidious microbes (fungus, virus, and mouth-loving bacterium) make all of us, still, potential clowns. We itch, we rub, we pick at the sores, we scratch. Inadvertently we spread them. We deliver *Candida* to its best food: mouth dribble. The fungus feeds on us, digests leftover

Figure 16.1 Zany, clown of his Mountebank in 1690 (Morley, 1880).

crumbs on our mouths. Of course the yeast has no mouth itself, yet even so *Candida* "gobbles up" our own chafed and peeling skin; seasoned by these moist tidbits *Candida* delights in dwelling at our lip cracks. I know a guy who is even *attracted* to girls with chapped lips. He says they look comely. He believes in chasing them on bicycles because there is no separation by metal and glass, as in cars, and yet the speed of the bike lends itself to more potential encounters. A pedaling womanizer on a mobile spade-shaped throne: sort of like the Joker in a pack of Bicycle cards. Little does he realize how fungal hunger drives "his" attraction.

THE FUTURE OF MAN

I respond to the computer voice emanating from the wheelchair amid dribbling tea. I sickly joke to the returning graduate student that our conversation has degenerated (we were just talking about the weather). He seems content now. But not long before the Cambridge University Professor of Physics Stephen Hawking had warned a full auditorium of the dangers of an airborne mutant of the AIDS virus. The charming, stuttering, brilliant cyborg of a futuristic oracle of a man had warned us that the viral scourges threaten to decimate humanity.

In *Galapagos*, a novel by Kurt Vonnegut, nearly all of mankind has been killed by nuclear holocaust and its pestilential aftermath of virus attack. Only a few isolated islanders survive the all-out war. These descendants of vacationing ecotourists mutate into blubbery pointy-headed primates, gentle, quiet souls who forget who their mothers are after two years. These former people enter the water splashlessly. After a few generations, only one trait remains that they share with us: like us, their landlubbing, big-headed walking ancestors, when they lie around to bask on the beach in the sun and one of them farts, all the others laugh.

But I'm not serious enough. How strangely hilarious it would be, in true Vonnegut fashion, if the microbes made it so itchy and painful that the human sex became a throwback, as it is in *Stentor coerleus* (figure 13.1, page 113). Cloning and in vitro fertilization will take care of reproduction. That and the microbes and birth control—and state mandates like in China and pollution-lowered sperm counts—will make intercourse all but obsolete pretty soon. The fungi will win: all artists, *Candida*-infested and virus-whipped, will have the smiles wiped off their faces. They will be as gloomy as those mass-produced, wood panelling-surrounded reproductions of paintings of clowns with bulbous red noses. Reproduction will be by government decree. Genetically engineered slave women, concubines of the state, will suffer artificial insemination by Bill Gates' sperm. Coitus interruptus at a global scale. For the good of the nerdocracy. The way of the *Stentor*.

CAMILLA'S MOUTH ERUPTIONS

Charles Bukowski called John Fante's *Ask the Dust* the greatest novel ever written. Late in the book Camilla Lopez, the waitress with whom Fante's semiautobiographical protagonist, Arturo Bandini, has fallen in love, goes out into the desert. The author, exasperated, throws his book, dedicated to her, toward her vanishing form. Arturo describes the scene just before he left: "It was like old times, our eyes springing at one another. But she was changed, she was thinner, and her face was unhealthy, with two eruptions at each end of her mouth. Polite smiles."

Fungi had the last laugh. The two eruptions marking Camilla's facial infirmity were outward signs of healthy, well-fed *Candida* colonies in their Edenic gardens. Those sad semicircles that make the tipsy clown's face look smiley from afar hide scaly skin. His sloughing mouth cells feed the happy fungi. He can't help it. A cold sore is a cold sore is a cold sore but on the corner of the mouth it is something else. I googled "cracked bleeding corners of mouth" and landed on the doors of dermatology in "perlèche," "angular cheilitis," and "candidiasis." Some people, desperate, endure it for years. Some have tried everything: balm and compress, silent lip movement, drinking from a straw. Desperate, some even superglue together their split lips, to no avail. The disease "scales." The fungi thrive on the scabs they produce themselves. The mouth reacts, generating delicious new dead and dissolving skin for the *Candida* cells to ingest. The worst the sufferer can do is what I did—drool while writing poetry and drinking microbrewery beers. Their frothy heads gush up to kiss me on my chapped lips. My slithery saliva became their nutritious drink, my ale, tasty, mighty tasty. The crust formed at my mouth's corners after I drank the beer. What else would a modern fool do?

VITAMIN-CRAVING YEAST

Cybelle, my girlfriend, requests carrot cake. She repeats herself. Carrot cake. She asks for it like a parrot, please, like a craving, the imperative of pregnant females. She still wants carrot cake. Maybe, I thought, she was inhabited by some microbial life-form that requires vitamin A. Carrot-cake desire had taken over her body. Because the Internet had told me that vitamin A overdose can make lips crack and a deficiency of vitamin B_{12} makes lips crack, too, we were between a rock and a hard place. Cybelle underestimated her twin's intelligence. Just because Marybelle has a Southern accent and Cybelle had purged herself of external signs of Florida did not mean they still aren't twins. In her eagerness to explain the cracking corners she had written down "candidiasis," or tried to: she spelled it as she heard it, with a K. This magnified the confusion. Bemoaning, we commiserated. We lacked the prescription that we could not pay any doctor to write. I sipped my microbrewed beer that I never should have ordered and decided to drink it via gulps. The split malt may have invigorated my mouth fungi and brought me into intimate contact with the medieval fool. "Be careful what you wish for," they say. For *angular cheilitis* and *perlèche* are just fancy names for that infamous white truth, ye olde yeast infection: *Candida* ("truth") *albicans* ("white"). It, and its fungal ilk, burgeoning populations of persistent pests, like it diaper damp and vagina moist; they are most partial to the slippery, lip-licked mouth and wet, food-friendly tongue. In so much of what I read they say *Candida* grow prodigiously at mouth corners. They inhabit folds and seams. Wrinkles, folds, and seams are but threads of fungal dreams. Those who treat *Candida* infections successfully, said the Internet Health Food Advice Service, go hog wild gorging themselves on yogurt and cottage cheese. They devour *Lactobacillus* ("acidophilus") to displace the yeast. They invite cheeky bacteria, like the resident *Staphylococcus*, to replace *Candida*. I applied cortisone, swallowed my big vitamin B_{12} pill. I washed my face with liquid soap after each meal (and most snacks). I applied Tinactin and generic Monistat (the girl cream) gently to my lips in a thin veneer. It got so bad my house no longer smelled like beer. Slowly the cracks receded as the fungus left me alone. I ate and smiled now, more often than not, without pain. My eyes no longer bulged. Here and there I even ventured to laugh. It was comic relief. I laughed at clowns, at large mouths with makeup circles at their edges. Their swollen red-painted lips hiding candidiasis. They still suffered candidiasis even if, alas, they were only pictures. Meanwhile I don't have it anymore. HaHaHa!

The first clowns evolved, did they not, from medieval fools and face-painting mimes? For what did they do but laugh and play? When their sore mouths, stubborn, refused to heal I suspect they painted on the smiles that God had wiped off. To smile was so painful that makeup protected them from the need to ever flash a real smile again. They drank and frol-icked and, ignorant of fungi, viruses, bacteria, indeed all microbes, inad-vertently fed their persistent perlèche. They plied mouth-corner fungi colonies daily with dribble, sweets, and fruits. The wine and beer only bid the yeast to grow even more vigorously and cause more smile pain. The average hardworking clown woke up drunk or hungover in a pool of his drool. Paradise for *Candida*. Nightly the clowns made themselves up. Day in and day out, again and again they disguised their smile pain by phony made-up smiles. They covered up the *Candida* cakes on the corners of their poor, sad mouths. Henceforth whenever they could not smile because it hurt too much their caked-on makeup smiled for them. When they died, *Candida*, freed at last, grew all over their corpses. (Indeed, the ancient Egyptians embalmed their dead because if they did not, gut *Candida* would grow at death and rapidly decompose the inner body, laying the table for worms.) Starting from the corners of the clowning mouths *Candida*, with other microbes, converted their bodily remains to ashes and dust. Wrinkles, folds, and seams: all that was left of the clown body and the phony-smiling clowning face was mashed-up clown cells, the ideal food for fungi. Digested, like a good joke, we return to the soil. Fungi have the last laugh.

Chapter 16 Notes
1. Sonea and Mathieu, 2002.
2. Kendrick, 2001.

part four

gaea

Runners-up for mythological mascot of this section were Uranus, the sky god and son of Gaea, and Urania, the mystic nymph, or make that muse of astronomy, often depicted with a rod in one hand and a globe in the other. But Gaea won out. The goddess of the Earth, Gaea is the mother of the Titans, Furies, and Cyclopes. Among the Titans, a family of primordial gods, were Hyperion, father of the Sun, Moon, and dawn, and Cronos, leader of the pack of that first family of gods who, feeling suffocated by his parents' uxorious embrace, castrated his father only to be ousted later by his son Zeus. Gaea is also the etymological root of *ge*, or *geo*, meaning Earth, found in such scientific words as geometry, geology, and geography. It is also an alternative spelling for Gaia, the name suggested to atmospheric chemist James Lovelock by novelist neighbor William Golding as a colorful way of calling attention to the theory that life on the surface of our planet behaves as a single, self-regulating system. Lovelock, as a respected inventor, was hired in 1961 by the National Aeronautics and Space Administration (NASA) Jet Propulsion Laboratory (JPL) in California to devise an instrument that could detect life on Mars. He was struck that his colleagues were all designing miniature bacteriology labs that could sample the alien surface and provide medium for the growth of bacteria, if any. How could they be sure that Martian life would grow in the culture, or be anything like life on Earth? Challenged by his boss to propose an alternative to life so narrowly understood, Lovelock proposed that a more general system of detection would be one that looked for entropy reduction on the level of a whole planet—something, in other words, that would detect "departure from the expected equilibrium state of a dead planet."[1] In September 1965 he had an epiphany: planetary entropy could easily be found by measuring the chemistry of the atmosphere and making a few simple thermodynamic equations. In the small JPL office he shared with Carl Sagan results came to them from an infrared telescope study at the Pic du Midi observatory in France in 1965. The atmospheres of Mars and Venus were almost entirely carbon dioxide! This implied that our planetary neighbors' atmospheres were close to chemical equilibrium and therefore lifeless. On the basis of geological evidence and the fossil record Earth's air, by contrast, was stable but away from equilibrium; it was

> dynamically stable like our blood plasma. . . . It then occurred to me that life must have been regulating the composition of the air for most of its existence. . . . When I returned to England a few days later it was to my home in the village of Bowerchalke in Wiltshire. A

near neighbor and friend was the author William Golding. He asked about my latest encounters with NASA and when I told him my idea he grew excited and said, "You must give it a proper name—I suggest you call your hypothesis Gaia, after the ancient Greek goddess of the Earth."[2]

Nowhere is poet T. S. Eliot's famous line "We shall not cease from exploration, and the end of all our exploring will be to arrive where we started and know the place for the first time" more apt than in its application to what we have learned about Earth in our attempt to find life elsewhere. The search for extraterrestrial life in our solar system and beyond has come up empty-handed, except perhaps for our remote sensing machines and instruments, which may be regarded as extensions of Earth life.

With respect to Lovelock's awesome discovery of Earth's life as a single, global life-form, it is perhaps ironic that the public is far more aware of the romantic but failed search for life elsewhere than it is of the startling discoveries of Gaia's global physiology and terrestrial intelligence.

Our tendency is to ignore the beating of our hearts until the irregularity demands our attention. We breathe bountiful oxygen in clean air at least once per several seconds without a care or second thought. When thirsty we drink cool, clear water. As long as air, water, food, and warmth are provided, we give these life essentials very little consideration. But who provides us with these uncelebrated requisites of happy lives? Money, though a powerful symbol, is only green linen, inedible paper, or scratchy numbers on a bank account sheet. Our parents and employers are not providers fulfilling the absolute necessity for air, water, food, and warmth. The only mother who can ensure our continued life is the bountiful third planet from the Sun in our minor spiral-armed galaxy—minor, that is, in the universe, but crucial to us. We call her Gaea or Gaia or Biosphere 1 or Planet Earth but she ought to be renamed Planet Water! She, our

providence and source, is the reason, in all her ambiva-
lences, complexities, and four-thousand-million-year history, that we, all thirty million or so species of the five
kinds of life on Earth, have subsisted and persisted, and
may yet have a future.[3]

This last section leads us to a greater awareness of our
whole blue-and-green planet, and to our debt not to each
other but to the biosphere. It should help our apostasy.
We authors try to lead you to return from the insane
anthropocentrism of the "civilized," monotheistic religious fantasies of the past few thousand years to the
healthier, out-of-doors Greek imagination that preceded
the "modern" aberration. Indeed, the combination of
genuine insights wrested from scientific inquiry with the
retention of the greatest achievements of the human
world from the coast of China to Australia, from Siberia
to Tierra del Fuego, in the end must be celebrated in its
proper place within a far greater planetary context.

Notes
1. Lovelock, 1979.
2. Lovelock, 1988.
3. Lovelock, 2006.

The Atmosphere, Gaia's Circulatory System

LYNN MARGULIS AND JAMES E. LOVELOCK

Originally, professional science was published in Latin. But there were some important exceptions. Scientists sometimes go right to the people. Galileo's defense of Copernicus's Sun-centered solar system, *Dialogue Concerning the Two Chief World Systems* (discussed also in chapter 20), was published in popular form, in Italian. Darwin's *Origin of Species*, published of course in English, was not just a technical but a popular book. Published first by Stewart Brand, founder of the *Whole Earth Catalog*, the following original, accessible essay is arguably the classic explanation of the Gaia hypothesis. It was first published not in a technical journal but in a popular magazine, *CoEvolution Quarterly*. "What do you care what other people think," Brand asked, "if this Gaia stuff is really science?"

We see Earth's atmosphere from a new viewpoint: as an integral, regulated, and necessary part of the biosphere. In 1664 Sachs von Lewenheimb, a champion of William Harvey, used the analogy shown in figure 17.1 to illustrate the concept of the circulation of blood. Apparently the idea that water lost to the heavens is eventually returned to Earth was so acceptable in von Lewenheimb's time that Harvey's theory was strengthened by the analogy.

The subtitle of the dissertation (which addresses itself to the famous anatomist Thomas Bartholinus) explains that it deals with the analogies between the circular motion of the water from and back to the sea, on the one hand, and that of the blood from and back to the heart, on the other. This motion is "circular" not because it describes the geometrical figure of a circle, but because it reverts to its point of departure. The Earth resembles the human body in that, like the latter, it is pervaded by canals and

Figure 17.1 Frontispiece to Sachs von Lewenheimb, 1664, *Oceanus Macromicrocosmicus.* According to W. Pagel, this illustration stresses the analogies between the circulation of the blood and the circulation of water.[2] From the original treatise in the Wellcome Library, London, courtesy of the trustees, with permission.

harbors an internal fire. The sea lets water rise by evaporation and return in the form of rain, whereby the rivers and subterranean waters are nourished and these finally return the same water to the sea. The latter thereby acts not unlike the heart, from which the blood goes out to the organs, starting on its way attenuated by the influx of heat and "perfected" in the "workshops" of the organs; finally, after its absorption and assimilation by the organs, its residue is drawn back into the heart in order to be attenuated again, just as the waters are diluted by joining the sea.

Three and a half centuries later, with the circulation of blood a universally accepted fact, we find it expedient to revive von Lewenheimb's analogy, this time to illustrate our concept of the atmosphere as circulatory system of the biosphere. This new way of viewing Earth's atmosphere has been called the Gaia hypothesis.[3] The term "Gaia" is from the Greek for "Mother Earth," and it implies that certain aspects of Earth's atmosphere—temperature, composition, oxidation reduction state, and acidity—form a homeostatic system, and that these properties are themselves products of evolution.[4]

Many articles and books [5] [6] [7] give one the impression that fluid dynamics, radiation chemistry, and industrial pollution are the major factors determining the properties of the atmosphere. The Gaia hypothesis contends that biological gas exchange processes, especially processes involving microorganisms, are also major factors. The human impact on the atmosphere may have been overestimated. Humans are only one of some three million species on Earth, all of which exchange gas and most of which exchange gas with the atmosphere. Humans have been around for only a few million years, while microorganisms have existed for thousands of millions of years. The atmosphere is probably not so much the product of humans as of the several billion smaller organisms living in every pail of rich soil or water.

It seems to us that early-twentieth-century nonmicrobiological analysis of Earth's lower atmosphere will one day be considered as ignorant as early-nineteenth-century nonmicrobiological analysis of fermentation or disease is today. "There is a great difference between research in the laboratory and studies of the Earth and planets. In the laboratory the scientist can perform controlled experiments, each carefully designed to answer questions of his own choosing. Except in minor respects, however, the Earth and planets are too large for controlled experimentation. All we can do is observe what happens naturally in terms of the laws of physics and chemistry," wrote Goody and Walker in their introduction to atmospheric science.[8]

We agree that the laws of physics and chemistry are basic to the understanding of atmospheric phenomena but insist that the laws of biology must be considered as well. It is our contention that the paucity of overall understanding of certain aspects of the atmosphere, especially composition and temperature, is due to too narrow a paradigm: the idea that the atmosphere is an inert part of the inorganic environment and therefore amenable to methods of study that involve only physics and chemistry.

In this chapter we explore what is perhaps a more realistic view: that the atmosphere is a nonliving, actively regulated part of the biosphere. In our model atmospheric temperature and composition are regulated with respect to certain biologically critical substances: hydrogen ions, molecular oxygen, nitrogen and its compounds, sulfur and its compounds, and some others, whose abundance and distribution in the atmosphere are presumed to be under biological control. Biological gas exchange processes, thought to be involved in possible control mechanisms, are discussed elsewhere.[4] The purpose of this chapter is simply to present our reasons for believing the atmosphere is actively controlled.

Traditional atmospheric studies have left us with some strange anomalies. The atmosphere is an extremely complex blanket of gas in contact with the oceans, lakes, and rivers (the hydrosphere) and the rocky lithosphere. It has a mass of about 5.3×10^{21} grams. (The mass of the oceans, the other major fluid on Earth's surface, is almost a thousand times heavier, being about 1.4×10^{24} grams.) Because the atmospheric mass corresponds to less than a millionth of the mass of Earth as a whole, one would expect small changes in the composition of the solid earth to cause large changes in the composition of the atmosphere. Yet even in the face of a large number of potential perturbations, the atmosphere seems to have remained dynamically constant over long periods of time.

Many facts about the atmosphere are known—its composition, its temperature and pressure profiles, certain interactions with incoming solar radiation, and the like.[8] Some of these are shown in tables 17.1 and 17.2. However, as the efficacy of long-range weather forecasting attests, there is no consistent model of the atmosphere that can be used for the purpose of prediction. Earth's atmosphere defies simple description. From the point of view of chemistry, it sustains such remarkable disequilibrium that Sagan[9] was prompted to remark that given the temperature, pressure, and amount of oxygen in the atmosphere, "one can calculate what the thermodynamic equilibrium abundance of methane ought to be . . . the answer turns out to be less than 1 part in 10^{36}. This then is a discrepancy of at least 30 orders of magnitude and cannot be dismissed lightly."

Table 17.2 shows that given the quantity of oxygen in the atmosphere, not only the major gases such as nitrogen and methane but also the minor atmospheric components are far more abundant than they ought to be according to equilibrium chemistry. Even though the minor constituents differ greatly in relative abundance, they sustain very large fluxes, comparable with those of the major constituents. Earth's atmosphere is certainly not at all what one would expect from a planet interpolated between Mars and Venus. It has too little CO_2 and too much oxygen gas, and it is

Table 17.1 Reactive Gases in the Atmosphere

Gas	Concentration in parts per million	How much of the gas (in billions of tons per year) comes from Inorganic sources (volcanoes, etc.)?	Biological sources? Gaian* / Human	Residence time**	Where does the gas principally come from?
Nitrogen (N_2)	790,000	0.001	1 / 0	1–10 million years	Bacteria from dissolved nitrate in soil
Oxygen (O_2)	210,000	0.00016	110 / 0	1,000 years	Cyanobacteria, algae, green plants; given off in photosynthesis
Carbon dioxide (CO_2)	320	0.01	140 / 16	2–5 years	Respiration, combustion
Methane (CH_4)	1.5	0	2 / 0	7 years	Methanogenic archeabacteria
Nitrous oxide (N_2O)	0.3	Less than 0.01	0.6 / 0	10 years	Bacteria and animals
Carbon monoxide (CO)	0.08	Less than 0.001	1.5 / 0.15	Few months	Methane oxidation (methane from bacteria)
Ammonia (NH_3)	0.006	0	1.5 / 0	Week	Bacteria, animals, and fungi
Hydrocarbons $(CH_2)n$	0.001	0	0.2 / 0.2	Hours	Green plants, industry
Methyl iodide (CH_3I)	0.000001	0	0.03 / 0	Hours	Marine algae
Hydrogen (H_2)	0.0000005	0	? / ?	2 years	Bacteria, methane oxidation, fermentation
Methyl chloride (CH_3Cl)	0.00000000114	0	? / ?	?	Algae?

*Gaian sources are nonanthropogenic biological sources, this table is abbreviated.

**Residence time is the time it takes for the concentration of gas to fall to 1/e or 37 percent of its value. It may be thought of as turnover time.

? = quantity unknown. See Lovelock & Margulis, 1974a, b and Smil, 2002 for details.

too warm. We believe the Gaia hypothesis provides the new approach that is needed to account for these anomalies.

A new framework for scientific thought is justified if it guarantees new observations and experiments. The recognition that blood in mammals circulates in a closed, regulated system gave rise to meaningful scientific questions such as: How is blood pH kept constant? By what mechanism is the temperature of mammalian blood regulated around its set point? What is the purpose of bicarbonate ion in the blood? What is the role of fibrinogen? If the blood were simply an inert environment (as the atmosphere is presently viewed), such questions would seem irrelevant and never be asked at all.

Table 17.2 Composition of the Atmosphere: Gases in Disequilibrium

Gas	Abundance	Flux (moles/yr X 10^{13})	Disequilibrium Factor	Oxygen Used Up in the Oxidation of These Gases (moles/yr X 10^{13})	Source of Gas (% Contribution) by	
					Abiological Processes	Biological Processes Human / Gaian*
Nitrogen	78%	3.6	10^{10}	11	0.001	0 / >99
Methane	1.5 ppm	6.0	10^{30}	12	0	0 / 100
Hydrogen	0.5 ppm	4.4	10^{30}	2.2	?	0 / ?
Nitrous oxide	0.3 ppm	1.4	10^{13}	3.5	0.02	0 / >99
Carbon monoxide	0.08 ppm	2.7	10^{30}	1.4	0.001	10 / 90
Ammonia	0.01 ppm	8.8	10^{30}	3.8	0	0 / 100

*Gaian sources are nonanthropogenic biological sources.

? = quantity unknown; ppm = parts per million

Let us consider another analogy. Bees have been known to regulate hive temperatures during midwinter at about 31°C, approximately 59°C above ambient.[10] Under threat of desiccation they also maintain high humidities. While the air in the hive is not alive, it maintains an enormous disequilibrium due to the expenditure of energy by the living insects—ultimately, of course, solar energy. How is the hive temperature maintained? How does the architecture of the hive aid to reduce desiccation? How does the behavior of the worker bees alter temperature? These are all legitimate scientific questions generated by the circulatory system concept.

The Gaia hypothesis of the atmosphere as a circulatory system raises comparable and useful scientific questions and suggests experiments that based on the old paradigm would never be asked. For example, how is the pH of the atmosphere kept neutral or slightly alkaline? By what means has the mean midlatitude temperature remained constant (not deviated more than 15°C) for the past one billion years? Why are 0.5 x 10^9 tons of nitrous oxide (N_2O) released into the atmosphere by organisms? Why is about 2 x 10^9 tons of biogenic methane pumped into the atmosphere each year (representing nearly ten percent of the total terrestrial photosynthate)? What are the absolute limits on the control mechanisms, that is, how much perturbation (emanations of sulfur oxides, chlorinated compounds, and/or carbon monoxide; alterations in solar luminosity; and so forth) can the atmosphere regulatory system tolerate before all its feedback mechanisms fail?

Table 17.3 Critical Biological Elements That May Be Naturally Limiting

Major Elements	Use in Biological Systems	Possible Form of Fluid Transport
C (carbon)	All organic compounds	CO_2; food; organic compounds in solution; biological volatiles; carbonate, bicarbonate, etc.; usually not limiting
N (nitrogen)	All proteins and nucleic acids	N_2, N_2O, O_3, NO_2 (often limiting)
O, H (oxygen, hydrogen)	H_2O in high concentration, H in organic compounds for all organisms	Rivers, oceans, lakes
S (sulfur)	Nearly all proteins (cysteine, methionine, etc.); key coenzymes	Dimethyl sulfide, dimethyl sulfoxide, carbonyl sulfide
P (phosphorus)	All nucleic acids; adenosine triphosphate	Unknown (biological volatiles? spores? birds? migrating salmon?)
Na, Ca, Mg, K (sodium, calcium, magnesium, potassium)	Membrane and macro-molecular function	Usually not limiting, except in certain terrestrial habitats[11]
Trace Elements		
I (iodine)	Limited to certain animals (e.g., thyroxine), algae	Methyl iodide
Se (selenium)	Enzymes of fermenting bacteria (production of ammonia, hydrogen) animals[12]	Unknown (dimethyl selenide?)
Mo (molybdenum)	Nitrogen-fixation enzymes of bacteria, including cyanobacteria and many other bacteria (e.g., *Clostridium*) that convert CO_2 gas into organic compounds.	Active site of enzymes

The Gaia approach to atmospheric homeostasis has also led to a number of observations that otherwise would not have been made. For example, an oceanic search was undertaken for volatile compounds containing elements that are limiting to life on the land, and large quantities of methyl iodide and dimethyl sulfide were in fact observed by Lovelock and his colleagues.

Given the Gaia hypothesis, one deduces that all the major biological elements (table 17.3) either must be not limiting to organisms (in the sense that they are always readily available in some useful chemical form) or must be cycled through the fluids on the surface of the earth in time periods that are short relative to geological processes The cycling times must be short because biological growth is based on continual cell division, which requires the doubling of cell masses in periods of time that are generally less than months, and typically are days or hours. On lifeless planets there is no particular reason to expect this phenomenon of atmospheric cycling, nor on Earth is it expected that gases of elements that do not enter metabolism as either metabolites or poisons will cycle rapidly; for example, based on the Gaia hypothesis, nickel, chromium, strontium, rubidium, lithium, barium, and titanium will not cycle, but cobalt, vanadium, selenium, molybdenum, iodine, and magnesium might.[13] Because biological solutions to problems tend to be varied, redundant, and complex, it is likely that all of the mechanisms of atmospheric homeostasis will involve complex feedback loops. Because, for example, no volatile form of phosphorus has ever been found in the atmosphere, and because this element is present in the nucleic acids of all organisms, we are considering the possibility that the volatile form of phosphorus at present is totally "biological particulate." Figures 17.2 and 17.3 rather fancifully compare Earth's atmosphere at present with what it might be if life were suddenly wiped out.

Ironically, it is Earth's history, with its extensive sedimentary record, fraught, as it is, with uncertainties in interpretation, that might provide the most convincing proof for the existence of continued biological modulation. If one accepts the current theories of stellar evolution, the Sun, being a typical star of the main sequence, has substantially increased its output of energy since Earth was formed some 4.5 billion years ago. Some estimates for the increase in solar luminosity over the history of Earth are as much as 100 percent; most astronomers apparently accept an increase of at least 25 percent over 4.5 billion years.[14] Extrapolating from the current atmosphere, given solar radiation output and radiative surface properties of the planet, it can be concluded that until about two billion years ago either the atmosphere was different (for example, contained more ammonia) or Earth was frozen. The most likely hypothesis is that Earth's atmosphere contained up to about one part in 10^5 ammonia, a good infrared absorber.[15] Other potential greenhouse gases apparently will not compensate for the expected lowered temperature because they do not have the appropriate absorption spectra or are required in far too great a quantity to be considered reasonable.[15] (There are good arguments for the rapid photodestruction of any atmospheric ammonia.[16]) However, it has been argued that ammonia is required for the origin of life,[17] and there is good evidence for the presence of fossil microbial life in the earliest sedimentary rocks (3.4 billion years ago).[18] There is no geological evidence that since the beginning of Earth's stable crust the entire planet has ever frozen solid or that the oceans were volatilized, suggesting that the temperature at the surface has always been maintained between the freezing

Figure 17.2 Earth's atmosphere with life: examples of major volatiles. The following compounds and spores are depicted; it is left to the reader to identify them. • Spores of: ferns, club mosses, zygomycotes, ascomycotes, basidiomycotes, slime molds, bacteria. All contain nucleic acids and other organic phosphates, amino acids, and lipids, and therefore generate "atmospheric volatiles." • Animal products: butyl mercaptan. Plant products: myoporum, catnip (nepetalactone), eugenol, geraniol, pinene, isothiocyanate (mustard), disparlure • Mainly microbial products: PAN (paroxacetyl nitrate), dimethyl sulfide, dimethyl sulfoxide. Gases: nitrogen, oxygen, methane, carbon monoxide, carbon dioxide, ammonia. Painting by Laszlo Meszoly.

and the boiling points of water. The fossil record suggests that, from an astronomical point of view, conditions have been moderate enough for organisms to tolerate, and the biosphere has been in continuous existence for over three billion years[18] (Cloud 1968).[19] At least during the familiar Phanerozoic eon (the last 541 hundred million years of Earth history, for which an extensive fossil record is available), one can argue on paleontological grounds alone that through every era Earth has maintained tropical temperatures at some place on the surface and that the composition of the atmosphere, at least with respect to molecular oxygen, could not have deviated markedly. That is, there are no documented cases of any metazoans (animals, out of about three million) that can complete their life cycles in the total absence of O_2. All animals are composed of cells

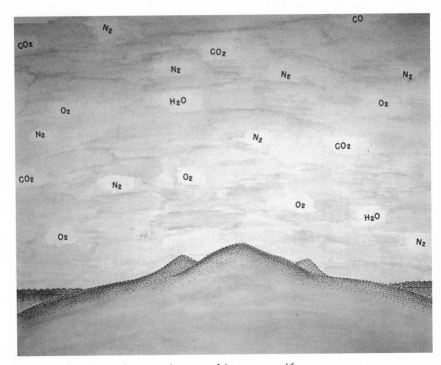

Figure 17.3 Earth's atmosphere now if there were no life. Painting by Laszlo Meszoly.

that divide by mitosis. In nucleated organisms that contain mitochondria, eg., all plants and animals, the mitotic cell division itself requires O_2. Thus it is highly unlikely that current concentrations of oxygen have fallen much below their present values in some hundreds of millions of years. By implication, oxygen and the gases listed in table 17.2 have been maintained at stable atmospheric concentrations for time periods that are very long relative to their residence times. (Residence time is the time it takes for the concentration of gas to fall to $1/e$ or 37 percent its value; it may be thought of as turnover time.) Furthermore, because concentrations of atmospheric oxygen only a few percent higher than ambient lead to spontaneous combustion of organic matter, including grasslands and forests, the most reasonable assumption is that the oxygen value of the atmosphere has remained relatively constant for quite long time periods.

How can these observations be consistently reconciled? How can we explain the simultaneous presence of gases that are extremely reactive with each other and unstable with respect to minerals in the crust and at the same time note that their residence times in the atmosphere are very short with respect to sediment-forming and mountain-building geological processes? In this respect table 17.3 can be instructive. One can see that even though absolute amounts of the gases vary over about three orders of magnitude, the fluxes are remarkably similar. These gases are produced and removed primarily by nonhuman biological processes (see table 17.1; also note 4). While the processes involved in atmospheric production and removal of reactive gases are not primarily dependent on human activity, for the most part they are not based on animal or plant processes either. The main organisms involved in gas exchange are the prokaryote microorganisms that dominated Earth as oxygen went from scarce to abundant.[19] Both now and in the preCambrian these rapidly growing and dividing masters of the microbiological world have made up in chemical complexity and metabolic virtuosity what they lack in complicated morphology. These organisms presumably played a role in biogeochemical processes of the past similar to the one they play today. There is direct fossil evidence for the continued existence of Precambrian microorganisms. That they have an ancient history can also be deduced from current studies of their physiology. Among hundreds of species of these prokaryotic microorganisms are many obligate anaerobes, that is, organisms poisoned by oxygen. (All organisms are poisoned by oxygen at concentrations above those to which they have become accustomed.) Hundreds of others are known that are either microaerophils (they live well in concentrations of oxygen less than ambient), facultative aerobes (they can switch their metabolism from oxygen requiring to oxygen nonrequiring), or aerotolerant (they are indifferent to the presence of oxygen or its absence since they don't use it in their metabolism).

As a group, the prokaryotic microbes show evidence that the production and release of molecular oxygen into the atmosphere was an extremely important environmental determinant in the evolution of many genera. Prokaryotic microbes (formerly known as the blue-green algae, or cyanobacteria) were almost certainly responsible for the original transition to the oxygen-gas-rich atmosphere that began about two thousand million years ago.

Figures 17.4 and 17.5 present scenes before and after the transition to oxidizing atmosphere, respectively. Figures 17.6 and 17.7 are reconstructions of anaerobic cycles corresponding to figures 17.4 and 17.5, respectively. Figure 17.4 attempts to reconstruct the scene as it might have looked 3,400 million years ago, admittedly in a rather geothermal area. Although no free oxygen (above that produced by photochemical processes and hydrogen loss) is available in the atmosphere, the scene is

Figure 17.4 Scene from a geothermal area on Earth during the Archean eon (about 3400 million years ago). Drawing by Laszlo Meszoly.

Figure 17.5 Scene from the Proterozoic eon (about two thousand million years ago). Drawing by Laszoy Meszoly.

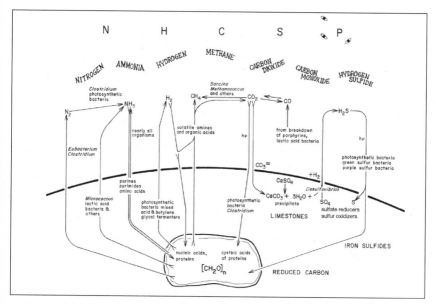

Figure 17.6 A reconstruction of metabolism in an anoxic world: 3,400 million years ago. (Genera of microorganisms catalyzing the reactions are underlined.) Drawing by Laszlo Meszoly.

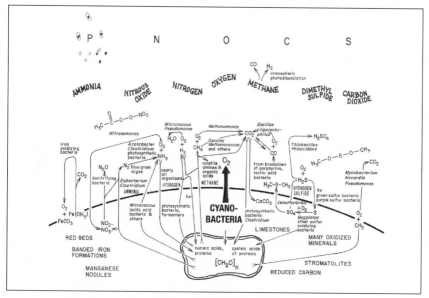

Figure 17.7 A reconstruction of microbial metabolism in an oxic world, begun about two thousand million years ago. Drawing by Laszlo Meszoly.

teeming with life—microbial life. For example, entire metabolic processes, as shown in figure 17.6, are available within the group of anaerobic prokaryotic microbes today. Because at the higher taxonomic levels (kingdoms and phyla) successful patterns, once evolved, tend not to become extinct,[21] it is likely that ancestors of present-day microbes were available to interact with atmospheric gases very early on the Archean Earth. Certainly life was very metabolically complex by the time the first stromatolitic rocks were deposited. With the evolution of oxygen-releasing metabolism by cyanobacteria came the stromatolites. These layered sediments are extremely common, especially in the late Precambrian.[22] With the stromatolites come other Precambrian evidence for the transition to the oxidizing atmosphere. By the middle Precambrian, about 2000 million years ago—the time at which the stromatolites and microfossils become increasingly abundant[23]—the scene might have looked like that in figure 17.5. The metabolic processes accompanying that scene are shown in figure 17.7. It is obvious that from among metabolic processes in prokaryotic microbes alone there are many that involve the exchange of atmospheric gases. Here we see how oxygen-handling metabolism was superimposed on an anoxic world, a concept that is consistent with the observation that reaction with molecular oxygen tends to be the final step in aerobic respiratory processes. All of the processes shown in these figures are known from current microorganisms (and, by definition, those that haven't become extinct are evolutionarily successful).

The fossil evidence, taken together, suggests that Earth's lower atmosphere has maintained remarkable constancy in the face of several enormous potential perturbations—at least the increase in solar luminosity and the transition to the oxygen-rich atmosphere. Earth's atmosphere maintains chemical disequilibria of many orders of magnitude containing gases that are produced in prodigious quantities and turned over rapidly. The temperature and composition seem to be set at values that are optimal for most of the biota. Furthermore, the biosphere has many potential methods for altering the temperature and composition of the atmosphere.[24] The biosphere has probably had these methods available almost since its inception more than three thousand million years ago. Is it not reasonable to assume that the atmosphere is maintained at an optimum by homeostasis and that this maintenance, because of the influx of solar energy, of course, is performed by the party with the vested interest: life itself?[25]

Chapter 17 Notes

1. Brand, author of several books, is also founder of the Point Foundation, cofounder of both the Global News Network and the Long Now Foundation. He influenced the course of cultural evolution when he helped all of us take hold of our own destinies with his magazine the *CoEvolution Quarterly*, which became the *Whole Earth Review*.

2. Pagel quotes Harvey himself as saying: "I began to think whether there might not be a motion as it were in a circle. Now this I afterwards found to be true; . . . which motion we may be allowed to call circular, in the same way as Aristotle says that the air and the rain emulate the circular motion of the superior bodies; for the moist earth, warmed by the sun evaporates; the vapors drawn upwards are condensed, and descending in the form of rain moisten the earth again; and by this arrangement are generations of living things produced. . . . And so in all likelihood, does it come to pass in the body, through the motion of the blood; the various parts are nourished, cherished, quickened by the warmer more perfect vaporous spiritous, and, as I may say, alimentive blood; which, on the contrary, in contact with these parts becomes cooled, coagulated, and, so to speak, effete; whence it returns to its sovereign, the heart, as if to its source, or to the inmost home of the body, there to recover its state of excellence of perfection." (Pagel, 1951.)

3. Lovelock, 1972; Lovelock, 1979.

4. Lovelock and Margulis, 1974 a and b.

5. Rasool, 1974.

6. Kellogg and Schneider, 1974.

7. Schneider et al., 2002.

8. Goody and Walker, 1972.

9. Sagan, C. 1970.

10. Wilson, 1970.

11. Botkin et al., 1973.

12. Stadtman, 1974.

13. Egami, 1974.

14. Oster, 1973.

15. Sagan, C. and Mullen, 1972.

16. Ferris and Nicodem, 1974.

17. Bada and Miller, 1968.

18. Barghoorn, 1971.

19. Cloud, 1968.

20. Gregory, 1973.

21. Simpson, 1960.

22. Awramik 1973.

23. Barghoorn and Tyler, 1965.

24. Schopf 1970.

25. For a summary of the status of the astounding four decades of work on evidence for life on the early Earth see Knoll, 2003.

Gaia and Philosophy

DORION SAGAN AND LYNN MARGULIS

The whole Earth-Gaia-goddess idea has many precedents. The human dilemma is whether our fundamental relationship is with Nature (Gaia in the Cosmos) or only with other people. The Daisyworld model invented by Lovelock and his close colleagues went far to begin to convince the world that Gaia is real, a testable scientific hypothesis with explanatory power. Scientists are a contentious lot, demanding evidence, not just belief. The evidence for global self-regulation was clear enough, but how it could be possible seems mysterious.

The Gaia hypothesis is a scientific view of life on Earth that represents one aspect of a new biological worldview. In philosophical terms this new worldview is more Aristotelian than Platonic. It is predicated on the earthly factual, not the ideal abstract, but there are some metaphysical connotations. The new biological worldview, and Gaia as a major part of it, embraces the circular logic of life and engineering systems, shunning the Greek-Western heritage of final syllogisms.

Gaia is a theory of the atmosphere and surface sediments of the planet Earth taken as a whole. The Gaia hypothesis in its most general form states that the temperature and composition of Earth's atmosphere are actively regulated by the sum of life on the planet—the biota. This regulation of Earth's surface by and for the biota has been in continuous existence since the earliest appearance of widespread life. The assurance of continued global habitability, according to the Gaian hypothesis, is not a matter merely of chance. The Gaian view of the atmosphere is a radical departure from the former scientific concept that life on Earth is surrounded by and adapts to an essentially static environment. That life interacts with and eventually becomes its own environment; that the atmosphere is an extension of the biosphere in nearly the same sense that the human mind is an extension of DNA; that life interacts with and controls physical attributes of Earth on a global scale—all these things resonate strongly with the ancient magico-religious sentiment that all is one. On a more practical plane, Gaia holds important implications not only for understanding life's past but for engineering its future.

The Gaia hypothesis, presently a concern only for certain interdisciplinarians, may someday provide a basis for a new ecology, and even become a household word. Already it is becoming the basis for a rich new worldview. Let us first examine the scientific basis for the hypothesis and then explore some of the metaphysical implications.

Innovated by the atmospheric chemist James Lovelock, supported by evolutionist Lynn Margulis, and named by novelist William Golding, the Gaia hypothesis states that the composition of all the reactive gases as well as the temperature of the lower atmosphere have remained relatively constant over eons. (An eon is approximately a billion years.) In spite of many external perturbations from the solar system in the last several eons, Earth's surface has remained habitable by many kinds of life. The Gaian idea is that life makes and remakes its own environment to a great extent; life reacts to global and cosmic crises, such as increasing radiation from the Sun or the appearance for the first time of oxygen in the atmosphere; and life dynamically responds to ensure its own preservation such that the crises are endured or negated. Both scientifically and philosophically, the Gaia hypothesis provides a clear and important theoretical window for what Lovelock calls "a new look at life on Earth."[1]

Astronomers generally agree that the sun's total luminosity (output of energy as light) has increased during the past four billion years. They infer from this that the mean temperature of Earth's surface ought to have risen correspondingly. But there is evidence from the fossil record of life that Earth's temperature has remained relatively stable.[2] The Gaia hypothesis recognizes this stability as a property of life on Earth's surface. We shall see how the hypothesis explains the regulation of temperature as one of many factors whose modulation may be attributed to Gaia. The temperature of the lower atmosphere is steered by life within bounds set by physical factors. With a simple model that applies cybernetic concepts to the growth, behavior, and diversity of populations of living organisms, Lovelock has most recently shown how, in principle, the intrinsic properties of life lead to active regulation of Earth's surface temperature. There is nothing mystical in the process at all. By examining in some detail the life of a mythical world containing only daisies (about which, more later), even skeptical readers can be convinced that it is theoretically possible for living, growing, responding communities of organisms to exert control over factors concerning their own survival. No unknown conscious forces need be invoked; temperature regulation becomes a consequence of the well-known properties of life's responsiveness and growth. In fact, perhaps the most striking philosophical conclusion is that the cybernetic control of Earth's surface by unintelligent organisms calls into question the alleged uniqueness of human intelligent consciousness.

In exploring the regulatory properties of living beings, it seems most likely that atmospheric regulation can be attributed to the combined metabolic and growth activities of organisms, especially of microbes. Microbes (or microorganisms) are those living beings seen only with a microscope. They display impressive capabilities for transforming the nitrogen-, sulfur-, and carbon-containing gases of the atmosphere. Animals and plants, on the other hand, show few such abilities. All or nearly all chemical transformations present in animals and plants were already widespread in microbes before animals and plants evolved. Until the development of Lovelock's Daisyworld, the discussion of control of atmospheric methane (a gas that indirectly affects temperature and is produced only by certain microbes, known as methanogenic bacteria) has provided the most detailed exposition of the maintenance of atmospheric temperature stability.[3] The concentration of water vapor in the air correlates with certain climatic features, including the temperature at Earth's surface. The details of the relationship between temperature and forest trees, which determines the production and transport of huge quantities of water in a process called evapotranspiration, was recently presented by meteorologists in a quantitative model.[4] Although meteorologists do not discuss their work in a Gaian context, they, and certain ecologists and geochemists, inadvertently provide Gaian examples. Indeed, as Hutchinson originally recognized when he described the geological consequences of feces,[5] and as books by Smil and others show, many observations concerning the effects of the biota in maintenance of the environment may be reinterpreted in a Gaian context.[6][7]

How can the gas composition and temperature of the atmosphere be actively regulated by organisms? Although willing to believe that atmospheric methane is of biological origin and that the process of evapotranspiration moves enormous quantities of water from the soil through trees into the atmosphere, several critics have rejected the Gaia hypothesis as such because they fail to see how the temperature and gas composition of an entire planetary surface could be regulated for thousands of millions of years by an evolving biota that lacks foresight or planning of any kind.

Primarily in response to these critics, Dr. Lovelock and his former graduate student Dr. Andrew Watson formulated a general model of temperature modulation by biota, to which they pleasantly refer as Daisyworld. Daisyworld uses surface temperature rather than gas composition to demonstrate the possible kinds of regulating mechanisms that are consistent with how populations of organisms behave. Daisy World exemplifies the kind of Gaian mechanisms we would expect to find, based as it is on an analogy between cybernetic systems and the growth properties of organisms. In an admittedly simplified fashion, it shows that temperature regulation can emerge as a logical consequence of life's well-known properties. These include potential for exponential growth, with growth rates varying with temperature such that the highest rate occurs at the optimal temperature for each population, decreasing around the optimum until growth is limited by extreme upper and lower temperatures. We will describe Daisyworld in detail shortly.

Some such model, explaining the regulation of surface temperature, is required to explain several observations. For example, the oldest rocks not metamorphosed by high temperatures and pressures, both from the Swaziland System of southern Africa[8][9] and from the Warrawoona Formation of western Australia,[10] contain evidence of early life. Both sedimentary sequences are over three billion years old. From over three thousand million years ago until the present, we have a continuous record of life on Earth, implying that the mean surface temperature has reached neither the boiling nor the freezing point of water. Given that an ice age involves less than a 10°C drop in mean midlatitude temperature and that even ice ages are relatively rare in the fossil record, the mean temperature at Earth's surface probably has stayed well within the range of 5° to 25°C during at least the last three billion years. Solar luminosity during the last four thousand million years is thought by many astronomers to have increased by at least 10 percent. Thus life on Earth seems to have acted as a global thermostat. Any current estimate for the increase of solar luminosity, which varies from less than 30 to more than 70 percent, does not alter the outcome of Daisyworld's conclusions. A relative increase of solar luminosity from values of 0.6 to 2.2 (its present value is 1.0) is consistent with Daisyworld assumptions because a range of values has been plotted by Lovelock and his collaborator Watson.[3]

Cybernetic systems, as is well known to science and engineering, are steered. They actively maintain specified variables at a constant in spite of perturbing influences. Such systems are said to be homeostatic if their variables, such as temperature, direction travelled, pressure, light intensity, and so forth, are regulated around a fixed set point. Examples of such set points might be 22°C for a room thermostat or 40 percent relative humidity for a room humidifier. If the set point itself is not constant but changes with time, it is called an operating point. Systems with operating points rather than set points are said to be homeorrhetic rather than homeostatic. Gaian regulatory systems, such as the embryological ones,[11] are more properly described as homeorrhetic rather than homeostatic. Fascinatingly enough, both homeorrhetic and homeostatic systems defy the most basic statutes of Western syllogistic thought, although not thought itself, because most people do not think syllogistically but in an associative fashion. For instance, if a person—surely a homeorrhetic entity—is hungry, he or she will eat. Thereupon hunger ceases. Put syllogistically, the sense of such a series becomes nullified: I am hungry; therefore I eat; therefore I am not hungry. The thesis leads to an antithesis without ever being synthetically resolved. This circular, tautological mode of operations is characteristic of cybernetic systems, including, of course, all organisms and organismic combinations. It is consonant with the emotive poetic power of contradictory statements, dichotomous personalities, and oxymoronic lyrics, such as references to a midnight sun.

Even minimal cybernetic systems have certain defining properties: a sensor, an input, a gain (the amount of amplification in the system), and an output. In order to achieve stability or to increase complexity, the output is compared with the set or operating point so that errors are corrected. Error correction means that the output must in some way feed back to the sensor so that the new input can compensate for the change in output. Positive or negative feedback, usually both, are involved in error correction. A first attempt to apply this sort of cybernetic analysis to the Gaia hypothesis involved development of the Daisyworld mathematical model, first by Lovelock and later by Watson and Lovelock together.[12] We turn now to the description of the model.

The Daisyworld model is used to demonstrate how planetary surface temperature might be regulated. It makes simple assumptions: The world's surface harbors a population of living organisms consisting only of dark and light daisies. These organisms always breed true. Each light daisy produces only light offspring daisies, and each dark daisy produces only its kind. Totally black daisies absorb all of the light coming on them from the sun, and totally white daisies reflect all of the light. The best temperatures for growth for both dark and light daisies are considered to be the same: no growth below 5°C, increasing growth as a function of temperature to an optimum at 20°C, and decreasing growth rate above the optimum to 40°C, at which temperature all growth ceases.

At lower temperatures darker daisies are assumed to absorb more heat and thus to grow more rapidly in their local area than lighter daisies. At higher temperatures lighter daisies reflect and thus lose more heat, leading to a greater rate of growth in their local area. The details have been published in technical journals and clearly summarized for a general audience. The graphs generated by models using these assumptions show that dark and light daisy life can, because of growth and interaction with light, influence the temperature of the planet's surface on a global scale. What is remarkable about the various forms of Lovelock and Watson's model is that the amplification properties of the rapid growth of organisms (here daisies) under changing temperatures are enough in themselves to provide the beginning of a mechanism for global thermal homeorrhesis, a phenomenon that some would rather see credited only to a mysterious life force. In general, in these models an increase in diversity of organisms, such as a greater difference between the lightness and darkness of the daisies, leads to an increase in regulatory ability as well as an increase in total population size.

Daisyworld is only a mathematical model. Even with its oversimplification, however, the Daisyworld model shows quite clearly that temperature homeorrhesis of the biosphere is not something that is too mysterious to have a mechanism.[13] By implication it suggests that other observed anomalies, such as the near-constant salinity of the oceans over vast periods of time and the coexistence of chemically reactive gases in the atmosphere, may have solutions that actively involve life-forms. The radical insight delivered by Daisyworld is that global homeorrhesis is in principle possible without the introduction of any but well-known tenets of biology. The Gaian system does not have to plan in advance or be foresighted in any way in order to show homeorrhetic tendencies. A biological system acting cybernetically gives the impression of teleology. If only the results and not the feedback processes were stated, it would look as if the organisms had conspired to ensure their own survival.

The Gaia hypothesis says, in essence, that the entire Earth functions as a massive machine or responsive organism. While many ancient and folk beliefs have often expressed similar sentiments, Lovelock's modern formulation is alluring because it is a modern amalgam of information derived from several different scientific disciplines. Perhaps the strongest single body of evidence for Gaia comes not from the evidence of thermal regulation that is modeled in Daisyworld but from Lovelock's own field, atmospheric chemistry.[14]

From a chemical point of view, Earth's atmosphere is anomalous. Not only major gases, such as nitrogen, but also minor gases, such as methane, ammonia, and carbon dioxide, are present at levels many orders of magnitude greater than they should be on a planet with 20 percent free oxygen in its atmosphere. It was this persistent overabundance of gases that react with oxygen, persisting in the presence of oxygen, that initially convinced Lovelock when he worked at the National Aeronautics and Space Administration (NASA) in the late 1960s and early '70s that it was not necessary for the *Viking* spacecraft to go to Mars to see whether life was there. Lovelock felt he could tell simply from the Martian atmosphere, an atmosphere consistent with the dicta of equilibrium chemistry, that life did not exist there.[15] Earth's atmosphere, in fact, is not at all what one would expect from a simple interpolation of the atmospheres of our neighboring planets, Mars and Venus. Mars and Venus have mostly carbon dioxide in their atmospheres and nearly no free oxygen, while on Earth the major atmospheric component is nitrogen, and breathable oxygen comprises a good one-fifth of the air.

Lovelock has compared Earth's atmosphere with life to the way the atmosphere would be without any life on Earth. A lifeless Earth would be cold, engulfed in carbon dioxide, and lacking in breathable oxygen. In a chemically stable system we would expect nitrogen and oxygen to react and form large quantities of poisonous nitrogen oxides as well as the soluble nitrate ion. The fact that gases unstable in each other's presence, such as oxygen, nitrogen, hydrogen, and methane, are maintained on Earth in huge quantities should persuade all rational thinkers to reexamine the scientific status quo taught in textbooks of a largely passive atmosphere that just happens, on chemical grounds, to contain violently reactive gases in an appropriate concentration for most of life.

In the Gaian theory of the atmosphere, life continually synthesizes and removes the gases necessary for its own survival. Life controls the composition of the reactive atmospheric gases. Mars and Venus, and the hypothetical dead Earth devoid of life, all have chemically stable atmospheres composed of over 95 percent carbon dioxide. Earth as we live on it, however, has only 0.03 percent of this stable gas in its atmosphere. The anomaly is largely due to one facet of Gaia's operations, namely, the process of photosynthesis. Bacteria, algae, and plants continuously remove carbon dioxide from the air via photosynthesis and incorporate the carbon from the gas into solid structures such as limestone reefs and eventually animal shells. Much of the carbon in the air as carbon dioxide becomes incorporated into organisms that are eventually buried. The bodies of deceased photosynthetic microbes and plants, as well as of all other living forms that consume photosynthetic organisms, are buried in soil in the form of carbon compounds of various kinds. By using solar energy to turn carbon dioxide into calcium carbonates or organic compounds of living organisms, and then dying, plants, photosynthetic bacteria, and algae have trapped and buried the once atmospheric carbon dioxide, which geochemists agree was the major gas in Earth's early atmosphere. If not for life, and Gaia's cyclical modus operandi, our Earth's atmosphere would be more like those of Venus and Mars. Carbon dioxide would be its major gas even now.

Microbes, the first forms of life to evolve, seem in fact to be at the very center of the Gaian phenomenon. Photosynthetic bacteria were burying carbon and releasing waste oxygen millions of years before the development of plants and animals. Methanogens and some sulfur-transforming bacteria, which do not tolerate any free oxygen, have been involved with the Gaian regulation of atmospheric gases from the very beginning. From a Gaian point of view animals, all of which are covered with and invaded by gas-exchanging microbes, may be simply a convenient way to distribute these microbes more numerously and evenly over the surface of the globe. Animals and even plants are latecomers to the Gaian scene. The earliest communities of organisms that removed atmospheric carbon dioxide on a large scale must have been microbes. In fact, we have a direct record of their activities in the form of fossils. These members of the ancient microbial world constructed complex microbial mats, some of which were preserved as stromatolites, layered rocks whose genesis both now and billions of years ago is due to microbial activities. Although such carbon-dioxide-removing communities of microbes still flourish today, they have been supplemented and camouflaged by more conspicuous communities of organisms such as forests and coral reefs.

To maintain temperature and gas composition at livable values, microbial life reacts to threats in a controlled, seemingly purposeful manner. For instance, if atmospheric oxygen were to decrease only a few percentage points, all animal life dependent on higher concentrations would perish. On the other hand, as Andrew Watson and his colleagues showed, increases in the level of atmospheric oxygen would lead to dangerous forest fires; small increases of oxygen would lead to forest fires even in soggy rain forests due to ignition by lightning. Thus the quantity of oxygen in the atmosphere must have remained relatively constant since the time that air-breathing animals have been living in forests—which has been over three hundred million years. Just as bees and termites control the temperature and humidity of the air in their hives and nests, so the biota somehow controls the concentration of oxygen and other gases in Earth's atmosphere.

It is this "somehow" that worries and infuriates some of the more traditional Darwinian biologists. The most serious general problems confronting widespread acceptance of the Gaia hypothesis are the perceived implications of foreknowledge and planning in Gaia's purported abilities to react to impending crisis and to ward off ecological doom. How can the struggling mass of genes inside the cells of organisms at Earth's surface know, ask these biologists, how to regulate macroconditions like global gas composition and temperature? The molecular biologist W. Ford Doolittle, for example, a man who because of his work is perhaps predisposed toward viewing evolution at smaller rather than larger levels, sees the Gaia hypothesis as untenable, a motherly theory of nature without a mechanism.[16]

Another scientist, the Oxford University evolutionist Richard Dawkins, is even more forceful in his rejection of the theory. Likening it to the "BBC Theorem" (a pejorative reference to the television documentary notion of nature as wonderful balance and harmony), Dawkins has extreme difficulty in imagining a realistic situation in which the Gaian mechanism for the perpetuation of life as a planetary phenomenon could ever have evolved. Dawkins[17] can conceive of the evolution of planetary homeorrhesis only in relation to interplanetary selection: "The universe would have to be full of dead planets whose homeostatic regulation systems had failed, with, dotted around, a handful of successful, well-regulated planets of which Earth is one."

These sound like forceful arguments, yet if the critics of Gaia cannot accept the notion of a planet as an amorphic but viable biological entity, they must have equal if not greater cause to dismiss the origin of life. Surely at one point in the history of Earth a single homeostatic bacterial cell existed that did not have to struggle with other cells in order to survive, because there were no other cells. The genesis of the first cell can no more be explained from a strict Darwinian standpoint of competition among selfish individuals than can the present regulation of the atmosphere. While the first cell and the present planet may both be correctly seen as individuals, they are equally alone, and as such they both fall outside the province of modern population genetics.

Nonetheless, Lovelock, a sensitive man with a deep sense of intellectual mischief, has answered his critics with one of their own favorite weapons: mathematical model making in the form of the aforementioned Daisyworld. Not believing that Earth's temperature and gases can be regulated with machinelike precision for billions of years, because organisms cannot possibly plan ahead, Lovelock's critics reject his personification of the planet into a conscious female entity named Gaia. Originally lacking an explicit mechanism and falling outside the major Darwinian paradigm of selfish individualism, it was and still sometimes is difficult for trained evolutionists to refrain from regarding Gaia as the latest deification of Earth by nature nuts. How can an entangled mass of disjointed struggling microbes, they ask, effect global concert of any kind, let alone to such an extent that we are permitted to think about Earth as a single organism? The answer, of course, is the kind of analysis explored in Daisyworld, and one still waits to see how those who accuse Lovelock of conscious mysticism and pop ecology will respond to it in all its mathematical intricacy.

Perhaps the greatest psychological stumbling block in the way of widespread scholarly acceptance of Gaia is the implicit shadow of doubt it throws over the concept of the uniqueness of humanity in nature. Gaia denies the sanctity of human attributes. If intricate planning, for instance, can be mimicked by cunning arrays of subvisible entities, what is so special about *Homo sapiens* and our most prized congenital possession, the human intellect? The Gaian answer to this is probably that nothing is so very special about the human species or mind. Indeed, recent research points suggestively to the possibility that the physical attributes and capacities of the brain may be a special case of symbiosis among modified bacteria.[18]

In real life, as opposed to Daisyworld, microbes, not daisies, play the crucial role in the continual production and control of rare and reactive compounds. Microbial growth is also partly responsible, through the production of heat-modulating gases as well as their production of variously colored surfaces, for the continuing thermostasis of Earth. Trees, involved in the production and elimination of carbon dioxide and water through their leaves, are also deeply involved, although they must be considered latecomers to the Gaian system. Evolutionarily, microbes were responsible for the establishment of the Gaian system. Insofar as larger forms of animal and plant life are essentially collections of interacting microbes,

Gaia may be thought of as still primarily a microbial phenomenon.[19] We human beings, made of microbes, are part of Gaia no less than our bones, made from the calcium from our cells, are part of ourselves.

In his article on classical views of Gaia, Hughes quoted the ancient Greek work *Economics* by Xenophon: "Earth is a goddess and teaches justice to those who can learn, for the better she is served, the more good things she gives in return." In the classical view, that is, of the Greek Gaia or Earth goddess and the Latin Tellus, Earth is a vast living organism. The Homeric hymn sings:

> *Gaia, mother of all, I sing oldest of gods.*
> *Firm of foundation, who feeds all creatures living*
> *on Earth.*
> *As many as move on the radiant land and swim in the sea*
> *And fly through the air—all these does she feed with her*
> *bounty.*
> *Mistress, from you come our fine children and bountiful harvests.*
> *Yours is the power to give mortals life and to take it away.*[20]

Although Gaia is reappearing in modern dress, the modern scientific formulation of the Gaian idea is quite different from the ancient one. Gaia is not the nurturing mother or fertility doll of the human race. Rather, human beings, in spite of our raging anthropocentrism, are relegated to a tiny and unessential part of the Gaian system. People, like brontosauruses and grasslands, are merely one of the many weedy components of an enormous living system dominated by microbes. Gaia has antecedents not only among the classical poets but also among scientists, most notably in the work of the Russian V. I. Vernadsky.[21 22] But Lovelock's Gaia hypothesis is a modern piece of science: it is subject to observational and experimental verification and modification.

There is something fresh, new, and yet mythologically appealing about Gaia, however. A scientific theory of an Earth that in some sense feels and responds is welcome. The Gaian blending of organisms and environment into one, wherein the atmosphere is an extension of the biosphere, is a modern rationalist formulation of an ancient intuitive sentiment. One implication is that there may be a strong biogeological precedent for the time-honored political and mystical goal of peaceful coexistence and world unity.

Contrary to possible first impressions, however, the Gaia hypothesis, especially in the hands of its innovator, does not protect all the moral sanctions of popular ecology. Lovelock himself is no admirer of most environmentalists. He expresses nothing but disdain for those technological critics he characterizes as misanthropes or Luddites, people who are "more concerned with destructive action than with constructive thought." He claims, "If by pollution we mean the dumping of waste matter there is indeed ample evidence that pollution is as natural to Gaia as is breathing to ourselves and most other animals."[1] We breathe oxygen, originally and essentially a microbial waste product. Lovelock believes that biological toxins are in the main more dangerous than technological ones, and he adds sardonically that they would probably be sold in health food stores if not for their toxicity. Yet there is no clear division between the technological and the biological. In the end, all technological toxins are natural, biological by-products that, though via human beings, are elements in the Gaian system. Similarly, legislation and lobbying attempts, such as the recent furor in the United States over the mismanagement of the Environmental Protection Agency, are nothing more or less than part of Gaian feedback cycles.

Ecologically speaking, the Gaia hypothesis hardly reserves a special place in the pantheon of life for human beings. Recently evolved, and therefore immature in a fundamental Gaian sense, human beings have only recently been integrated into the global biological scene. Our relationship with Gaia is still superficial. However, our ultimate potential as a nervous early warning system for Gaia remains unsurpassed. Deflecting oncoming asteroids into space and spearheading the colonization of life on other planets represent additions to the Gaian repertoire that our species must initiate. On the one hand, Gaia was an early and crucial development in the history of life's evolution. Without the Gaian environmental modulating system, life probably would not have persisted. Now, only by comprehending the intricacies of Gaia can we hope to discover how the biota has created and regulated the surface environment of the planet for the last three billion years. On the other hand, the full scientific exploration of Gaian control mechanisms is probably the surest single road leading to the successful implementation of self-supporting living habitats in space. If we are ever to engineer large space stations that replenish their own vital supplies, then we must study the natural technology of Gaia. Still more ambitiously, the terraformation of another planet, for example, Mars, so that it can actually support human beings living out in the open is a gigantic task and one that becomes thinkable only from the Gaian perspective.

In terms of the metaphysics of inner space, acceptance of the Gaian view leads almost precipitously to a change in philosophical perspective. As just one example, human artifacts, such as machines, pollution, and even works of art, are no longer seen as separate from the feedback processes of nature. Recovering from Copernican insult and Darwinian injury, anthropocentrism has been dealt yet another reeling blow by Gaia. This blow, however, should not send us into new depths of disillusion or existential despair. Quite the opposite: we should rejoice in the new truths of our essential belonging, our relative unimportance, and our complete dependence upon a biosphere that has always had a life entirely its own.[7]

Chapter 18 Notes

1. Lovelock, 1979.
2. Lovelock and Margulis, 1974.
3. Watson et al., 1978.
4. Shukla and Mintz, 1982.
5. Hutchinson, 1954.
6. Botkin and Keller, 1987.
7. Smil, 2003.
8. Schopf, 1983.
9. Walter, 1976.
10. Awramik et al., 1983.
11. Waddington, 1976.
12. Watson and Lovelock, 1983.
13. Lovelock, 1983b.
14. Lovelock is an inventor as well as a scientist. He devised the electron capture device, a sensor for gas chromatographs that detects freon and other halogenated compounds in concentrations of far less than one part per million in the air. Indeed, it was Lovelock's invention and observations that, in large part, sparked ecological concern about ozone depletion, ultraviolet-light-induced cancers, and general atmospheric catastrophe. See Lovelock, 2006.
15. Lovelock and Margulis, 1976; for a recent statement of this use of a thermodynamic/chemical strategy to seek life beyond the Earth, see K. H. Nealson's chapter 24 "Life is a Mistake" in *Mind, Life, and Universe* (2007) edited by Lynn Margulis and Eduardo Punset, pages 211–216.
16. Doolittle, 1981.
17. Dawkins, 1982.
18. Margulis and Sagan, 1997.
19. Kaveski et al., 1983.
20. Hughes, 1983.
21. Vernadsky, 1998.
22. Lapo, 1988.

The Global Sulfur Cycle
and *Emiliania huxleyi*

DORION SAGAN

Fewer have heard of the sulfur cycle than of the carbon or
nitrogen cycles of the chemical elements on Earth, but sulfur
is crucial to all cells and their proteins. As V. I. Vernadsky recognized, sulfur in its various chemical forms must continue
to flow or all life would perish.

Certain elements are planetary lifeblood. Like blood, they flow through
the biosphere in limited supply. The carbon, sulfur, nitrogen, phosphorus,
oxygen, and hydrogen that make up all organisms on Earth are not infinite. They must be continually redistributed, or cycled. Unlike an animal,
Earth has no heart pushing this global flow in a simple beat. Instead, the
planet lives on a complex of different forces all pulsing to a syncopated
rhythm. These forces include wind, daily sunlight and darkness, ocean
currents and tides, the erosion and surges of volcanoes and mountain
building, the separation and collision of continents, and the incessant
motions of living beings.

Although it became more feasible with satellite measurements in the late
twentieth century, tracking global element cycles is still a Herculean task. But
the National Aeronautics and Space Administration (NASA), which is used
to studying planets as whole entities, is turning its resources toward Earth.
Every other year since 1980 a NASA-supported group called Planetary
Biology and Microbial Ecology (PBME) has brought together academics,
researchers, and space scientists to discuss the connections between life and
the elements it needs to sustain itself. In 1980 the group looked at many elements. The focus was carbon in 1982. In 1984 PBME–NASA tried to determine sulfur's elusive path through the "veins" of the world. Nitrogen was to
be the next mystery element but with a two-decade hiatus, there has been a
long lag. Now, John F. Stolz, a former PBME student and now professor of
biology at Duquesne University in Pittsburgh, plans a new PBME to research

not only the nitrogen cycle but all the great element cycles. The "extremophiles," microbes, master cycles of carbon, nitrogen, and sulfur will be studied on location in Yellowstone National Park in collaboration with Montana State University in 2009, if all goes well.

Many of the major transformations that keep elements accessible to life transpire in hot springs, salt flats, and deeply textured sediments called microbial mats. To investigate these environments, PBME participants met in San Jose, California, in 1984, and explored the San Francisco Baylands, Alum Rock Wildlife Refuge, and Big Soda Lake in the two-casino town of Fallon, Nevada. Here scientists tried to piece together the puzzle of sulfur-using microbes and the global sulfur cycle. The program's long-term goal is to blend space technology and microbiology to come up with a map, as it were, of global metabolism. But in the short term, the scientists must trek amid a stench resembling rotten eggs and cabbages, braving pools of mud and suspiciously colored gunk.

An analogy for the collective work of PBME–NASA may be found in the early anatomical studies of the Renaissance artist and scientist Leonardo da Vinci. PBME is also on the vanguard of exploration, uncovering the mechanics of the biosphere. But whereas Da Vinci cut open bodies and looked inside them to be able to draw and abstract about the human body, today's interdisciplinarians—environmentalists, petroleum geologists, microbial ecologists, soil scientists, oceanographers, and atmospheric scientists—study small samples of the biosphere with the constant knowledge that the larger system they are part of can be viewed at large, imaged in near entirety from space.

Just as metabolism is the complex of chemical activities that maintains the structure of organisms and their component cells, so the metabolic activities of all organisms sharing planet Earth are so intimately linked that they form a sort of giant metabolism. Sulfur, part of this Earth-wide metabolism, is found in the proteins of all organisms and is therefore required for all growth.

The element exists in both hydrogen-rich forms and in highly oxidized forms. Chemical reactions, from oxidized to hydrogen-rich compounds and vice versa, yield energy. Life mediates sulfur and other elements through such chemical reactions, building up cell material or releasing energy for physiological processes.

Many bacteria, such as *Desulfovibrio*, *Desulfuromonas*, and *Desulfutomaculum*, turn oxidized sulfates ($SO_4^=$) and sulfur into hydrogen-rich sulfides. Sulfides, often in the form of gaseous hydrogen sulfide (H_2S),

Figure 19.1 *Emiliania* from space. Coccolithophorid bloom seen from space. The white is the land of the southwest coast of the British Isles. The coccolithophorids produce chlorophyll, which accounts for the dark green of the sea; they are also a major producer of dimethyl sulfide, a gas extremely important in the global sulfur cycle. This image helps us see how a phenomenon on the microorganism level could be discovered by planetary observations from space. Scientists have only recently realized that the dimethyl sulfide so important to the global sulfur cycle comes largely from these algae. (NASA)

are then used as an energy source for other bacteria, such as *Beggiatoa*. *Beggiatoa* need oxygen to get energy from oxidizing sulfide, but sulfide can even be oxidized under conditions where there is no gaseous oxygen by bacteria such as *Chromatium*, which use the oxygen in their cells to effect the transformation in reactions that may have originated on primordial Earth.

Microbes are key to the concept of element circulation, and they can be important in depositing major sulfur-containing minerals, such as the gypsum ($CaSO_4 \cdot 2H_2O$) found in salt flats. As William Holser, of the University of Oregon, told the PBME group, even pyrite (FeS_2), the familiar iron sulfide mineral known as fool's gold, ultimately depends on bacterial alteration of marine sulfate for its formation in sediments. If such mineral deposits depend on and, in a real sense, are part of life, then why are they considered static, inanimate, and nonliving? In fact, it may be better to look at such deposits as part of a global skeleton or storage system, one that is drawn upon by life in the way a pregnant woman draws upon the calcium of her bones to feed her fetus.

Prior to the 1980s few suspected that there was much sulfur in the atmosphere, except for the oxidized sulfur compounds from coal mining and the like. But atmospheric dimethyl sulfide $(CH_3)_2S$, a recent focus of attention, exemplifies a change of perception in interdisciplinary global studies toward seeing life and the environment, biology, and geochemistry as inextricably bound.

SULFUR IN THE AIR

Dimethyl sulfide, for example, which makes the sea smell like the sea, was caught ten years ago carrying huge amounts of sulfur from the ocean to the atmosphere. These sulfurous migrations, like most chemistry on Earth, are largely dependent on life.

Meinrat Andreae, at the Department of Oceanography at Florida State University, now at Mainz, Germany, discovered a correlation between the population density of marine algae such as *Phaeocystis* and *Emiliania* and the buildup of dimethyl sulfide (figures 19.1 and 19.2). Some of this gas, which brings so much sulfur up from seawater into the air, is produced by *Phaeocystis poucheti*. This obscure alga apparently uses the precursor to atmospheric dimethyl sulfide as an osmolyte, a compound that regulates intracellular salt concentration. For oceanic plankton exposed to the vicissitudes of changing salt concentrations, osmolytes are hot commodities.

Osmolytes can also be based on nitrogen compounds, but sulfur osmolytes are probably common in ocean-faring organisms, as well as being major sources of atmospheric sulfur gases.

Not all atmospheric sulfur gases are produced by microbes, of course. As New York City commuters from northern New Jersey know only too well, the activities of people also make significant contributions to the sulfur cycle. All factories and automobiles emit at least some sulfur dioxide (SO_2) when sulfur-bearing fossil fuels such as gasoline, coal, and oil are burned. Catalyzed by light, sulfur dioxide and oxygen react in the atmosphere to form sulfur trioxide (SO_3). Sulfur trioxide combines in water to make sulfate droplets that become the sulfuric acid (H_2SO_4) that, swept by winds from such places as the heavily industrialized Ohio Valley, falls as acid rain in New York and New England.

But Andreae said that nonhuman biological processes emit sulfur gases at rates at least comparable with the sulfur dioxide flux from fossil-fuel burning. In 1985 the amount of sulfur dioxide given off from the biota to

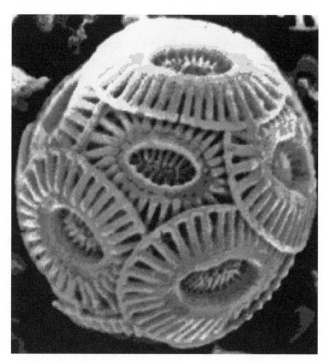

Figure 19.2 *Emiliania huxleyi,* a coccolithophorid. The carbonate scales, "buttons" (coccoliths) in their immense quantities lead to the white color visible off the south coast of the United Kingdom.

the air was, he said, on the order of a hundred trillion grams. By far the most important processes of the biogenic (natural as opposed to industrial) release of sulfur gases to the atmosphere is the chemical transformation of ocean sulfate into other forms of sulfur compounds by bacteria.

The incorporation of sulfate and organic sulfur compounds by algae and plants is a second immensely important sulfur transformation that occurs on a planetary scale. Indeed, British atmospheric chemist James Lovelock suggests that the quantity of such sulfur compounds—those produced by organisms other than humans and released into the atmosphere—may in fact be far greater than those produced by factories, power stations, and automobiles.

As part of conventional oceanography, environmental sulfur dioxide readings have traditionally been taken at sea. Andreae, Lovelock, and others feel, however, that estimates of sulfur gas production over land are probably wildly inaccurate, leading researchers to overestimate the volume of sulfur produced by industry. Part of the problem of determining the sulfur cycle is the difficulty of measurements: sulfur gases can vary by several orders of magnitude over a period of hours at one spot on the coast. Most of the acid rain precursors have been measured on land in the context of some specific, local pollution problem rather than in the context of a total understanding of Earth's atmosphere.

IN THE RAIN

Robert Fuller, of the Department of Civil Engineering at Syracuse University, reminds us that acid rain is only one in a suite of factors determining the acidity of lake water. A lake is frequently a small part of a much larger watershed, where water interacts with vegetation, soil, and the underlying rocks. Watershed characteristics, such as the presence of coniferous vegetation, high levels of soil organic carbon, shallow soils, an inability to adsorb and immobilize sulfate, and low levels of exchangeable and weatherable basic cations, are all factors that can predispose an ecosystem to transfer atmospheric acidity to surface waters. The alkalinity of the bedrock is involved, as well. As an example, neighboring lakes receiving acid rain in upstate New York have been found to have significantly different acidities. But these lakes, beneath the same sky, receive the same amounts of sulfuric acid in their rain.

These observations don't excuse the high sulfur emissions by industry. But they do show that the measured acidity in a lake does not depend only on the quantity of acid in the rain. Most of the furor about high levels of atmospheric sulfur and acid lakes comes from foresters, farmers, and fly-casters. Lakes have even been declared dead because of their relatively high concentrations of sulfuric acid. But not only trees, fish, and forest mammals are affected by acid rain.

In acidified lakes, as in sulfide-rich waters, there are many organisms that positively thrive. Indeed, unusually lush algal and bacterial growth may even identify a lake's acidity. Animals may flourish in high-acid lakes too: while trout are decimated or even totally killed off in very acidic lakes, causing indisputable economic hardship to people who depend on fishing, certain species of crayfish crawl about and reproduce to high population densities unperturbed. The types of bacteria that form coatings and mats, especially along the bottom of acid-rich lakes, are organisms with multibillion-year histories. These prolific microbes must have been involved in the formation of the earliest sulfur cycles.

PBME participants believe the major environmental sulfur transformations are fundamentally biochemical processes that evolved inside bacterial cells. Bacteria coevolved with the earliest biosphere, their remains existing as fossils in some of the oldest unmetamorphosed rocks. Although evidence for sulfur reduction—bacterial conversion of sulfate into sulfur and sulfide—appears in the fossil record only after the appearance of photosynthesis, there is some consensus that sulfate-reducing bacteria evolved before and paved the way for the development of photosynthesis.

A FREE LUNCH

Early in the history of life, anaerobic bacteria partook of the free lunch of energy-rich chemicals left over from the production of the so-called prebiotic soup. Yet soon after, we suggest, they evolved a more efficient way of deriving energy.

By diverting high-energy electron carriers away from the process of fermentation, some kinds of anaerobic bacteria evolved the ability to breathe the common oceanic ion sulfate. The ability to breathe sulfate and to use it instead of prebiotically produced complex organic sulfur compounds, such as the amino acids methionine and cysteine, gave such early anaerobic bacteria an evolutionary advantage: the more complete oxidation of organic matter provided them with additional energy.

To reduce carbon dioxide from the air into the hydrogen-rich carbon compounds of cells, microbes needed a source of electrons. An excellent early source of electrons was gaseous hydrogen, which was far more plentiful in the early solar system than it is today. As time went on the sun's high-energy radiation and Earth's weak gravitational field caused hydrogen to escape into space. Most early hydrogen was eventually lost from Earth's atmosphere, but hydrogen sulfide, a gas emitted from Earth's interior through hydrothermal vents, volcanoes, and sulfur hot springs, was still plentiful. Bacteria grappled with this for their electrons instead.

Today the green and purple sulfur bacteria still use hydrogen sulfide as their electron donor in photosynthesis. When cyanobacteria (formerly called blue-green algae) began using the hydrogen of water as an electron donor, the global sulfur cycle, along with the other major chemical cycles of the biosphere, changed forever. The use of water rather than hydrogen sulfide led to new waste products.

In the early days photosynthesis was largely dependent on a steady source of hydrogen sulfide, and the gas was converted into yellow sulfur deposits on the ground or into globules in the water that were later oxidized to make ocean sulfate. But now, as water replaced hydrogen sulfide as the largest reserve of electrons for photosynthesis, oxygen began to build up in the air. As the oxygen-producing cyanobacteria spread, the entire planet underwent dramatic oxidation. By 1,800 million years ago, during the Proterozoic eon, hydrogen-rich iron, uranium, and sulfur-bearing minerals at Earth's outer crust practically disappeared, replaced by oxygen-rich forms. But the biochemical legacy of the early hydrogen-rich environment was simultaneously preserved in the form of life, making Earth an astronomical oddity.

Because of life's oxygen waste, Earth underwent many new energizing and energy-releasing reactions, which in turn were exploited by life. The transition to an oxygenic biosphere had many literally Earth-changing consequences, among which was the banishment of some bacteria, those that had previously flourished at the surface, to a new subsurface realm of marine muds and warm geysers. To this day such oxygen-shunning bacteria make up the lower layers of the flat purple and green communities known as microbial mats.

Yehuda Cohen, of the Hebrew University in Jerusalem, introduced the use of microelectrodes as a means of measuring minute concentrations of oxygen, hydrogen, and sulfide in microbial mats. The new technique, first applied to microbial ecology by N. P. Revsbech, of Aarhus University in Denmark, allows detailed vigils over chemical transformations at Earth's surface. Microelectrode work ("physiology") coupled with ultrastructural study ("anatomy") show that the sedimentary layers of organisms that form these microbial mats are distinct in much the same way that skin, fat, and muscle tissue are composed of differentiated flattened masses of animal cells.

Certain chemical conditions, oxygen and sulfide concentrations, and levels of light penetration typify each layer, but differences in these variables can cause major changes in community interaction, and changes in community interaction in turn can feed back into changes in the variables. Cohen's team examined community relations among microbes in the salt flats near Leslie Salt Company in Newark, California. They looked at the surface and subsurface microbes in the sulfur springs of Alum Rock State Park too. In both of these locations the tiny millimeter-thick region that separates cyanobacteria from the sulfur bacteria rises slightly during the night and descends correspondingly during the day. At night there is no photosynthesis to produce the oxygen lethal to sulfur users, and so the microelectrodes detect increased levels of hydrogen sulfide closer to the

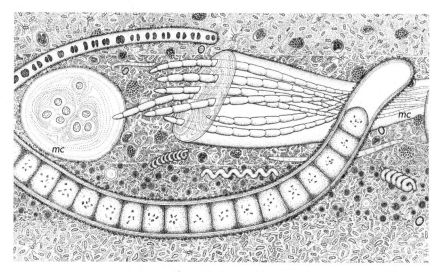

Figure 19.3 Cutaway view of microbial mat at Bido Salina, Matanzas, Cuba.
Drawing by Christie Lyons.

surface. Like the chest of a sleeper, the chemical boundary moves. Each day the hydrogen-sulfide/oxygen interface rises; each night it falls.

Some bacteria living in this zone are very versatile, as they must be to cope with potentially poisonous concentrations of both hydrogen sulfide and oxygen. *Oscillatoria limnetica*, for example, uses either hydrogen sulfide or hydrogen from water during photosynthesis. The cosmopolitan microbe *Microcoleus chthonoplastes* has a chameleon physiology. This cyanobacterium that looks like microscopic bundles of insulated wire (*mc* in figure 19.3), sometimes lives like an ancient bacterium, never producing any oxygen. Other times it performs the oxygen-producing photosynthesis typical of plants, but under concentrations of sulfide that would poison plants, animals, algae, and even other bacteria. It seems plausible that such versatility comes from a time when the gas mixture of Earth's atmosphere was changing from an oxygen-poor to an oxygen-rich one.

CHANGING NEIGHBORHOODS

The daily movement of the boundary layer between oxygen and sulfide may at times reflect not changes in the composition of communities of organisms so much as flexibility in the metabolism of those organisms. The surfaces of marshes, salt ponds, and muds bombarded by light from above and permeated with gas-containing fluids from below present a vast array of energy sources and opportunities. Those organisms able to vary their metabolic repertoire, to complement or enhance the metabolism of others, or just generally to be at home in the melee of deposition and gas exchange around the surface zone of sunlight grow like weeds. And they make the greatest contributions to the sulfur cycle.

To follow globally roaming elements whose territory is the entire surface of the globe is not simple. Sulfur, like any element important to life, has multiple guises and creates a web of activity crossing subtly between animate and inanimate realms. The marriage of microbial and planetary studies is an ambitious new enterprise. It may, like Leonardo Da Vinci, be ahead of its time.

The late Robert M. Garrels, of the University of South Florida, a PBME faculty participant and expert on element cycling, waxed ironic over global metabolism. Although Garrels took the idea of a giant circulatory and living system seriously, as shown by his remark that "Earth's surface environment can be regarded as a dynamic system protected against perturbations by effective feedback mechanisms," he also had a warning: "We all build more and more complicated geochemical models until no one understands anyone else's model. The only thing we do know is that our own is wrong."

But should we then give up trying to understand the global cycling of elements so important to life on Earth? Not necessarily. Garrels explained, "The chief purpose of our models is not to be right or wrong but to give us a place to store our data."

While NASA's life sciences program has been expanding in recent years to include Earth as a planet to be viewed from space and compared with its lifeless neighbors Mars and Venus, microbiology, geology, and chemistry have simultaneously become more circumscribed and circumspect in their university settings. This peculiarity of scientific history has led to an academic struggle, a hybrid sometimes called microbiogeochemistry. We will have to wait to see where this chimeric discipline leads. We still do not know whether it will ever be able to discover the metabolic workings of Earth or to plot the movement of the elements as gracefully as Da Vinci drew a man. Yet microbiogeochemistry, perhaps better called "geophysiology," is more than on the verge of a renaissance,[1] it has come of age.[2]

Chapter 19 Notes

1. Lovelock, 2006.
2. Margulis and Dolan, 2002.

Descartes, Dualism, and Beyond

DORION SAGAN, LYNN MARGULIS, AND RICARDO GUERRERO

Perhaps no invisible influence of unstated assumptions over intellectual discourse, including science, is greater than that of Cartesian dualism. Science is supposedly "objective," but all evidence and observations are collected by humans, peculiar mammals whose rampant dichotomization—good/bad, primitive/advanced, body/mind—may reflect the bilateral partitioning of the brain into two hemispheres. In the real world things, in particular the split between the body and mind, are not so divided.

The brilliant French Catholic mathematician René Descartes (1596–1650) inaugurated the mechanistic dichotomy with his declaration of a universal split between *res extensa*, the determined material reality of nature, and *res cogitans*, the free-thinking reality of people and God. Only humans, Descartes argued, partake of God to the extent that they have souls. Animals, though they seem to feel pain, are in fact soulless machines: "We are so accustomed to persuade ourselves that the brute beasts feel as we do that it is difficult for us to rid ourselves of this opinion. But if we were as accustomed to seeing automata which imitate perfectly all those of our actions which they can imitate, and to taking them for automata only, we should have no doubt at all that the irrational animals are automatons."[1]

Although Descartes' presentation of the universe as a vast mechanism led to an expansion of scientific investigation, the acceptance of the Cartesian mechanistic universe also had negative implications. On the authority of Descartes, scientists nailed live animals to boards without remorse to illustrate the facts of anatomy and physiology. Rationalized as unfeeling and inanimate, nature, in the wake of Descartes, was analyzed without fear of trespass. Nature, including the mechanical, automata-like "lower" life-forms, could now be experimented on with impunity. In short, Descartes' philosophy provided a formal justification—a Cartesian license—to investigate virtually everything in an effort to discover the mechanism by which God had "built" the phenomenal world.

By splitting reality into human consciousness and an unfeeling, objective exterior (or in his terms *extensive*) world that could be measured
mathematically, Descartes paved the way for a scientific investigation of
nature constructed according to the mathematical laws of God. "God sets
up laws in nature just as a king sets up laws in his kingdom," he wrote.[2]
The Cartesian license separated matter from form, body from soul, and
outward, spatially extended nature from inner awareness. Matter, body,
and nature could—unlike thought or feeling—be measured, compared,
and thus ultimately understood by mathematical laws.

This Cartesian license permitted the human intellect, through science, to
enter a thousand different realms, from the gigantic to the subvisible. The
once divine was now open to scientific exploration. Optical instruments
focused on snowflakes and peppercorns or pointed at the pockmarked
whiteness of the side-lit moon. Atoms were investigated by chemical combination and physical acceleration. X-rays imaged bones. Radioactive elements
clocked the internal metabolism of the human body. Eventually aeronautical
engineers even appropriated the seemingly God-given power to fly.

Investigation of the formerly divine realm yielded impressive scientific
results. Scientists, perusing nature and not books, returned the Bible and
the classics to their dusty shelves. There is a biographical anecdote, perhaps apocryphal, that when Descartes was asked in his urban domicile
about the location of his library, he pointed to a dissected calf he had been
examining and said, "On top of those books." Scientists began to study
the world, "written," as Galileo had put it even prior to Descartes, "in a
great book which is always open before our eyes."[3] Galileo had paid
dearly for his inquisitive temperament. As a quantitative mechanicist,
measurer of falling bodies, and discoverer of the moons of Jupiter and the
rotation of Earth,[4] it was Galileo who had cleared the trail for curious
successors such as Descartes, Newton, and the "Prince of Astronomy,"
William Herschel (1738–1822), who confirmed that the Milky Way is a
spiral-shaped object formed by distribution of its component stars.[5]

A defier of potent philosophers and Christian theologians, Galileo pro-
voked the ire of Church authorities. He was, at age fifty-eight, brought
before the Inquisition and charged with heresy. Galileo recanted his earlier
claims that were so at variance with official Church doctrine. He
"admitted" that Earth is at the center of the universe. Warned against further heresy, Galileo, who became a prisoner in his own country home, was
condemned to three years of weekly psalm recitations. Indeed, his thoughts
were censured for nearly two hundred years; until 1838 Galileo's

immensely popular masterpiece, *Dialogue Concerning the Two Chief World Systems*, was banned. With horror, Pope Urban VIII had recognized himself in Galileo's imagined character "Simplicio." Correctly believing that he had been mocked, it was Urban who began the censorship.

If Galileo had worked under the Cartesian license he would have fared better. When in 1633 the devout Descartes learned of Galileo's condemnation, he abandoned work on a manuscript that supported a heliocentric rather than an Earth-centered world. Impelled to integrate science into religion, Descartes gave great impetus to modern practices of investigation by doubting everything but the existence of his own doubting mind. Bodies, he held, were clocklike mechanisms, created by a Creator. The body is connected to the mind, he wrote, via the pineal gland, a pea-sized structure at the base of the brain, known at that time in the seventeenth century only in humans. The pineal acted, Descartes suggested, as a valve through which God was connected to the free human soul.

The Cartesian license still rallies scientists to study a universe wide open for investigation. But the "fine print"—to extend the metaphor—of this great card of admission into once forbidden realms ironically vouchsafes the same repressive, religion-based legacy it was designed to combat. Generating the mechanistic body is the conscious human mind in its deistic incarnation as the mind of God. This vitalistic residue of primordial consciousness remains the ghost within the machine of would-be wholly materialistic modern science. The Cartesian license still contains in its metaphorical fine print the following assumption: the universe is mechanical and is set up according to immutable laws by God. But neither the human exception to the predetermined laws of nature nor the metaphysical assumption of divine mechanism is science. Cartesian philosophy is more imbued with the historical presuppositions of western European culture than with the pure objectivity it touts.

Ultimately, we suggest, the Cartesian license proves to be a kind of forgery. After three centuries of implicit renewal, the permit is still valid, even though the fine print, worn off or ignored, is barely visible. Yet the fine print exempting humans and making machinate the "objective world" is no more peripheral to the Cartesian license than is the surgeon general's warning on a box of cigarettes. The raison d'être, the rational basis that authorized scientists to follow the spirit of Descartes to proceed with their work and to receive the blessings of society, including the Church, are already implicit in Descartes' license. For many centuries the Judeo-Christian religions had placed "man," man as "made in God's image," high on the ladder of being. People, in the cultural mind of the literate world, are situated perhaps a little lower than the angels but certainly above all the rest of life.

While Descartes cogitated, Europe remained under the rule of royalty. The king and the lord, representing the power and order of God, reigned supreme. But licensed Cartesians—medical men, explorers, alchemists—soon entered the realms which formerly they had been forbidden to enter for fear of transgressing the sacred.

Scientific revelation of mechanism, part of the new audacity of inquiry, helped unsettle European monarchy. If the universe, made by God, is a giant automaton that works itself, why should people obey any king or lord whose power, God-given in the feudal system of medieval Christianity, no longer derived from heavenly decree? Many began to take seriously what they took to be the implications of liberating free inquiry. High-born Frenchman Donatien Alphonse François Sade, as the infamous Marquis de Sade, for example, keenly wrote about and lived his conviction that the religious basis for morality had vanished. If Nature were a self-perpetuating machine and no longer a purveyor of divine authority, then why did the outrageous acts that he performed matter at all? All was, at best, the morally neutral turning of wheels in a vast, more lifelike than living, automatic mechanism.[6]

In 1776 the British colonists in North America broke free from transatlantic rule. Independence from the burdens of taxes and royalty was proclaimed. In 1789 the French Revolution deposed the king and stripped the lords and ladies of their powers. Irreverent Voltaire claimed that if God did not exist it would be necessary to invent Him. A century later the German philologist and nihilistic aesthetician Friedrich Nietzsche declared outright that God is dead. He defined philosophy as the unfettered love of knowledge and the philosopher as he before whom everyone quivers. "Philosophy," he wrote, is "a terrible explosive in the presence of which everything is in danger."[7]

England, too, was struck by the revolutionary spirit of the late eighteenth and early nineteenth centuries. Expansionist and socially moderate, however, the English, retaining their king and queen, perceived themselves a bastion of order in a world gone mad.

The Cartesian influence was profound. By the late nineteenth century, Western thought suffered a metaphysical reversal. The diminution of importance of the God-given human body and mind was more and more supported by the expanding, skeptical scientific worldview. Our prescientific ancestors tended to consider the universe and everything that moved to be alive. Beings were exempted from life only when they stopped moving, only when the spirit left them by the natural magic trick of death. But now things had changed: in the new scientific-mechanistic world of Galileo, Descartes, and Newton, the universe and all the beings in it were inanimate. The scientific puzzle moved from the mystery of death in a live cosmos to that of life in a dead one.

Inanimate matter had been rendered soulless and dead by the mechanists. Even animate matter was soulless and dead in the minds of strict Cartesians, who, with time, began losing their sway. But the universe is neither the dead mausoleum investigated by the Cartesian license nor an enchanted fairyland of invisible spirits.

We all, as citizens, scientists, scholars, or simply curious readers, are interested in life because we admire it from the inside. We feel life is something more than purely mechanical, and yet its freedom, if it exists, seems dubious to credit to a divine God. We do react to stimuli but we also seem to be able to think, to act, to choose. We seem far more than either Cartesian automata or entirely predictable Newtonian machines. Perhaps we are neither. But if we are more than Cartesian automata, so, après Darwin, must be the rest of life. Otherwise we risk a great inconsistency.

This dualistic cultural inheritance presents a continuing challenge to science. Given the limited legacy of Cartesian dualism (mind/body, spirit/matter, life/nonlife), it may not be surprising that two of the most profound twentieth-century rethinkers of life and its context share a biospheric perspective yet have diametrically opposed views. Russian scientist Vladimir Ivanovich Vernadsky (1863–1945) described organisms as he described minerals, calling them "living matter," whereas English scientist James E. Lovelock (b. 1920), our friend and colleague, has problematized Earth's surface in such a way that the entire biosphere, including rocks and air, may be regarded as alive.

Vernadsky portrayed living matter as a geological force—indeed, the greatest of all geological forces. Life moves and transforms matter across oceans and continents. Life, as flying phosphorus-rich seagulls, racing schools of mackerel, and sediment-churning polychaete worms, traverses the near-Earth environment, chemically transforming our planet's surface. Life, at the expense of the sun's energy, has been largely responsible for the great differences between the third planet and her solar-system neighbors, specifically, the unusual oxygen-rich and carbon-dioxide-poor atmosphere of Earth relative to those of Venus and Mars.

In a tradition begun by Christian Gotfried Ehrenberg (1795–1876), Alexander von Humboldt (1769–1859), and other serious explorers before him, Vernadsky described what Ehrenberg called the "everywhereness of life."[8] Living matter, he noticed, almost totally penetrated into, and consequently became involved in, superficially "inanimate" processes of weathering, water flow, and wind circulation. While his contemporaries spoke of the animal, vegetable, and mineral kingdoms, Vernadsky analyzed Earth's phenomena without labeling and classifying them into these categories. He eschewed preconceived notions of what was and was not alive. Perceiving life not as some abstract entity, with its philosophical, historical, and religious connotations, he referred only to "living matter." This freed him to combine as needed mineralogy, geology, and biology in a new discipline.

Impressed by the movement of machines in the World War I, what struck Vernadsky most was that the material of Earth's crust is packaged into myriad moving beings whose reproduction and growth depend on solar energy while they build and break down matter. Life, he saw, was a global phenomenon. Humans, for example, are accelerators of life's tendency to redistribute and concentrate the chemical elements of Earth: iron, aluminum, oxygen, hydrogen, nitrogen, carbon, sulfur, and phosphorus. Many other elements of Earth's crust are rapidly altered and mobilized by living beings, especially the two-legged, upright wanderers of our own species. People, he explained, have an amazing propensity to dig into, build up, move around, and in countless other ways alter the chemistry of Earth's surface. We, in Vernadsky's view, represent a new phase in biogeochemical evolution.[9]

Vernadsky contrasted gravity, which pulls material vertically toward the center of Earth, with life, which grows, runs, swims, and flies against the gravitational force. Life, challenging gravity, moves matter horizontally across the surface. Vernadsky detailed the structure and distribution of aluminosilicates in Earth's crust and was the first to recognize the importance of heat released from radioactivity to geological change.

But even a resolute materialist like Vernadsky found a place for mind. In Vernadsky's view a special thinking layer of organized matter, growing and changing Earth's surface, is associated with humans and technology. To describe it he adopted the term *nöosphere*, from the Greek *nöos*, "mind." The term *nöosphere* itself was introduced by Edouard Le Roy, of the Collège de France. Vernadsky met Le Roy through intellectual discussions in Paris in the 1920s; there he also met Pierre Teilhard de Chardin, the French paleontologist and Jesuit priest whose writings would later bring the idea of *nöosphere*—a conscious layer of life—to a wider audience. Teilhard's and Vernadsky's use of the term *nöosphere*, like their slants on evolution in general, differed. For Teilhard the nöosphere was the "human" planetary layer forming "outside and above the biosphere," while for Vernadsky the *nöosphere* referred to humanity and technology as an accelerating yet integral part of the planetary biosphere.

Vernadsky distinguished himself from other theorizers by his staunch refusal to erect a special category for life. For him life was far less a thing with properties than a happening, a process. Living beings in Vernadsky's writings are moving, chemically curious, but predictable forms of the common liquid mineral H_2O, which we call water. Animated water, life in all its wetness, displays a power of movement exceeding that of limestone, silicate, and even air. It shapes Earth's surface. Emphasizing the continuity of watery life and rocks, such as that evident in coal or fossil limestone reefs, Vernadsky developed the idea, later elaborated, that apparently inert geological strata are "traces of bygone biospheres."

Vernadsky and Lovelock, global scientists both but from distinct vantage points, have articulated ways in which life is far more than a Cartesian automaton, or any other sort of machine. The worldviews of both, complementary and complex, were constructed from the usual scientific observations of minutiae. Many eluded them both in spite of their keen powers of observation and sharply focused careers.

Consider this: when offered a variety of foodstuffs, bacteria, ciliates, mastigotes, and other swimming microbes make selections—they choose. Squirming forward on retractable pseudopods, *Amoeba proteus* finds *Tetrahymena* delectable but avoids *Copromonas*. A paramecium prefers to feed on small ciliates, but if starved for these and other protists, it reluctantly sweeps aeromonads and other bacteria into its cell mouth.

Although "merely" protoctists, foraminifera ("forams" for short) are one of the most diverse groups of fossil-forming small organisms. An astounding variety of magnificent shells are made by these complex single-celled beings, some forty thousand different species of which have evolved in the past 541 million years. Forams outside their shells resemble amoebae, with a network of long, thin, fusing and branching pseudopods. In certain forams, those called agglutinators, the shells are formed from handy starting materials from the seashore environment. Sand, chalk, sponge spicules, even other foram shells are patched together (agglutinated) to make the coverings. To appropriate their cell-shell homes, these forams

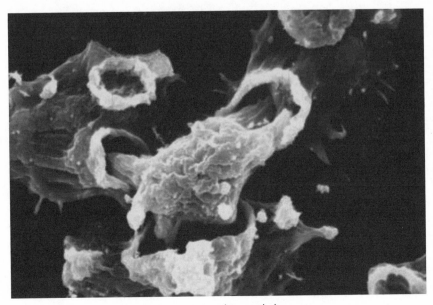

Figure 20.1 Amoeba cannibalism.

place available particles from their surroundings together with an organic cement. Experiments have shown, however, that when presented with a hodgepodge of different particles, foraminifera make distinct "choices" based on shape and size—selecting, for example, small black over larger red glass beads. Some will bridle at the term *choice*; however, there seems to be no reliable criterion for distinguishing between the preferential activities of these beings and ourselves. Without brains or hands, these protists pick the building materials from which to construct their body-homes.

Smaller still, and far simpler in cell organization, chemotactic bacteria can sense chemical differences. These little bodies, just two microns (two millionths of a meter) long, swim toward sugar and away from acid. A chemotactic bacterium, without a nose, of course, can "smell" a difference in chemical concentration that is a mere one part in ten thousand more concentrated at one end of its body than at the other.

Biochemist and former editor-in-chief of the leading scientific journal *Science*, Daniel Koshland, expressed the spiritual tendencies of the colon bacterium, *E. coli*, as follows:

"Choice," "discrimination," "memory," "learning," "instinct," "judgment," and "adaptation" are words we normally identify with higher neural processes. Yet, in a sense, a bacterium can be said to have each of these properties . . . it would be unwise to conclude that the analogies are only semantic since there seem to be underlying relationships in molecular mechanism and biological function. For example, learning in . . . [animal] species involves long-term events and complex interactions, but certainly induced enzyme formation must be considered one of the more likely molecular devices for fixing some neuronal connections and eliminating others. The difference between instinct and learning then becomes a matter of time scale, not of principle.[10]

Many organisms too small to be seen without a microscope sense and avoid heat and move toward or away from light. Certain bacteria even detect magnetic fields. Some harbor magnets aligned in rows along the length of their tiny, rod-shaped bodies.[11] That bacteria are simply machines, with no sensation or consciousness, seems no more likely than Descartes' claim that dogs suffer no pain. We reject the idea that microbes act without any feeling. Although possible, the idea is ultimately solipsistic. (Solipsism is the idea that everything in the world, including other people, is the projection of one's own imagination.) Cells, alive, act as if they have feelings. Indigestible mold spores and certain bacteria are rejected by protists. Others are greedily ingested. At even the most primordial level, living seems to entail sensation, choosing, mind.

For nineteenth-century men of science it was natural and expedient in the Cartesian tradition to invoke physical mechanisms to explain life. Life, as Newton's matter, consists of material bits that predictably respond to forces and obey natural laws. Like well-made clockwork, the world's mechanism was manufactured by the transcendent God, the creative God that constructed magnificent mathematical laws and then withdrew from his perfect and knowable creation.

Life, though, was not created in six days. Ushered in by the shocking contribution of Charles Darwin was the new view of evolution. God, if He existed, was Newton's God. No active interloper in human details, He was a geometer God who made the laws. Beneath the new mathematical God was the ancient residuum of the idea of a more active God.

The earlier view of life, the idea that life itself was evolving but only partially mechanical, was championed by Samuel Butler (1835–1902), an English novelist, painter, musician, and essayist whom Gregory Bateson called "Darwin's most able critic."[12] Butler took issue with the overly mechanistic views of Darwin.[13] He suggested no grand design in nature but recognized the continuity of life, to which he attributed millions of little purposes. Each purpose or objective was attributable to the cell or organism in its habitat.

To Newtonians, Darwinians, and others in the direct lineage of Descartes, choice or "free will" had been banished from a mechanistic universe. For Descartes, God, of course, has consciousness and people do as well, but only insofar as they communicate with God. When Darwin's painstaking work led to the conclusion that, like nonhuman life, people too had evolved (by the "mechanism" of natural selection), the consciousness that definitively separates Man from the Other suddenly became redundant. Butler, who argued against the special status of cogitating man, brought consciousness back into the discussion. He claimed that life is exuberant matter that chooses now and has chosen in the past. Over the eons choices made by some life-forms have produced more and different organisms, including the colonies of cells that stick together and become human individuals. Butler rejected a perfect immovable mathematical God; his deity is imperfect and dispersed. The properties of life, for Butler, lie in all life. "God" and life are one.

Butler's view that rejects any single, universal, omnipotent architect appeals to us. Life is too shoddy a production, both physically and morally, to have been designed by some austere flawless Master. And yet life is more impressive and less predictable than any object whose nature can be accounted for solely by "forces" acting on it deterministically. Butler's godlike qualities of life on Earth include neither omniscience nor omnipotence. Perhaps, though, an argument could be made for the omnipresence of Earthly life.

In the form of myriad cells, from luminescent bacterium to lily-hopping frog, life dwells everywhere throughout the surface of our third planet. All life is connected through Darwinian time and Vernadskian space. Evolution places us all in the stark but fascinating context of the cosmos. Although mystical powers may determine this cosmos, their existence is impossible to prove. The cosmos, more dazzling than any god of any particular religion, is enough for us. Life is existence's celebration. The features of purpose and determination that our culture tends to ascribe uniquely to people inhere intact in all of life. From life's minimal state as a tiny walled bacterial cell to its huge presence as a calf-nursing elephant or a montane rain forest, its exuberance and its sensible and sentient features apply to all of its forms.

Butler's theory intrigues us. We agree that mind and body are not separate but part of the unified, functioning whole. Life, sensitive from its onset, has been capable of choice, of decision, of sensing and thinking, from the beginning. Such "thoughts," both vague and clear, are physical. They are in the cells of our bodies and in those of other animals.

In comprehending these sentences, certain ink squiggles trigger associations, the electrochemical connections of the brain cells. Glucose is chemically altered by reaction of its components with oxygen, and its breakdown products, water and carbon dioxide, enter tiny blood vessels. Sodium and calcium ions, pumped out, traffic across a neuron's membranes. As you remember, nerve cells bolster their connections, new cell adhesion proteins form, and heat dissipates. Thought, like life, is matter and energy in flux; the body is its complement. Thinking and being are aspects of the same physical organization and its action.

If one accepts the fundamental continuity between body and mind, thought is essentially like all other physiology and behavior. Thinking, like excreting and ingesting, results from lively interactions of a being's chemistry. Even microbial "thinking" derives from cell hunger, movement, growth, association, programmed death, satisfaction, and other intrinsica of all life. Restrained but healthy former microbes find alliances to construct and behaviors to practice. If what is called "thought" results from such cell interactions, then perhaps communicating organisms, each themself thinking, can lead to a process greater than individual thought. This may be implicit in the Vernadskian notion of the nöosphere.

Two modern neuroscientists, Gerald Edelman (of the Scripps Institute in La Jolla, California) and William Calvin (of the University of Washington Medical School in Seattle), have each proffered concepts of mind. From Edelman's work and fertile imagination comes the phrase *neural Darwinism*. Our brains, both would agree, become minds as they develop by rules of natural selection.[14] This concept ultimately may provide a physiological basis for Butler's insights. In the developing brain of a mammalian fetus, some 10^{12} neurons each become connected with one another in 10^4 ways. These cell-to-cell adhesions at the surface membranes of nerve cells are called synaptic densities. As brains mature, over 90 percent of their cells die. By programmed death and predictable protein synthesis, connections selectively atrophy or hypertrophy. Neural selection against possibilities, always dynamic, leads to choice and learning as the remaining neuron interactions strengthen. Cell adhesion molecules synthesize and some new synaptic densities form and strengthen as nerve cells selectively adhere and as practice turns to habit. Selection is *against* most nerve cells and their connections, but it is nevertheless *for* a precious few of them. Of course, new work may reveal the physical basis of thought and imagination, but little doubt exists that selective cell death in a vast field of proliferating biochemical possibilities may apply to developing minds in the same manner it does to evolutionary change.[14]

Perhaps Descartes did not dare admit the celebratory sensuality of life's exuberance. He negated that the will to live and grow emanating from all live beings, human and nonhuman, is declared by their simple presence. He ignored the existence of nonhuman sensuality. His legacy of denial has led to mechanistic unstated assumptions. Nearly all our scientific colleagues still seek "mechanisms" to "explain" living matter, and they expect laws to emerge amenable to mathematical analysis. We demur; we should shed Descartes' legacy that surrounds us still and replace it with a deeper understanding of life's sentience. In Butler's terms, it is time to put the life back into biology.

It will cost our culture until we recover our senses[15] and return to the awareness that we must fully reject Cartesian anthropocentrism. We are interconnected not only to other people but to all other living beings on this planet's surface. The received view is that air travel, telephone lines, Internet connections, waterways, and fax machines connect only people. This view, symptomatic of residual Cartesian anthropocentrism, is biologically naive. In fact, such rapidly communicating methods connect, through us and others, all life. They link not only us but also our planet-mates. For inhabitants of the urban ecosystem the connections are obvious; whether or not we are conscious of others—cockroaches, sparrows, tomato plants, pigeons, and pubic lice—they clearly enjoy habitat expansion as we "develop" Earth for more people.

In retrospect, the Cartesian denial is exposed; we see Descartes' strategy as a Christian relic based on philosophical preconception rather than attentive observation. At this late date in our Western heritage, we can shed our Cartesian mechanistic legacy at no risk to our scientific credibility. Consistency precludes Cartesianism, or, indeed any kind of monotheism, Christian or not. Either we are like other live organisms in that both we and they exert choices, or both we and they are mechanistic, deterministic beings whose ability to choose behavior is essentially illusory. The middle ground is philosophical quicksand. The great majority of the inhabitants of this third planet in our solar system are not humans, nor have they ever been human. They enjoy and suffer all sorts of strife, social humiliation, joy, and victimization. They, too, choose intimacy or rejection. We feel that scientists and all others who continue to ignore the members of an estimated thirty million species, the other sentient beings, do so to their own great loss. Our planetmates whose existence Descartes and so many of his modern-day successors deny are communicants of the nonhuman splendor that, if we let them, can infuse our lives with delight and meaning.

Chapter 20 Notes

1. Jonas and Jonas, 2001.
2. Berman, 1989.
3. Jacob, 1973.
4. Simmons, 1996.
5. A large sign saying "Study Nature Not Books" attributed to nineteenth century polymath scientist Louis Agassiz decorates the library at the Marine Biological Laboratory in Woods Hole, Massachusetts.
6. Klossowski, 1991.
7. Wakeford and Walter, 1995.
8. Vernadsky, 1998.
9. Lapo, 1988.
10. Koshland, 1992. For microbial choice, also see chapter 5, pages 36–41.
11. Madigan et al., 1996.
12. Bateson, 1928.
13. For more on Butler and magnetotatic bacteria see chapter 9 "Sentient Symphony" in Margulis and Sagan, 2000, and chapter 10 in this book.
14. Edelman, 1985.
15. Abram, 1996.

What Narcissus Saw:
The Oceanic "Eye"

DORION SAGAN

When, without leaving Earth, we gaze upon Earth from space via satellite imaging technology, something strange is going on. We are, in a sense, seeing ourselves for the first time.

Sense-knowledge is the way the palm knows the elephant in the total pitch-dark. A palm can't know the whole animal at once. The Ocean has an eye. The foam-bubbles of phenomena see differently. We bump against each other, asleep in the bottom of our bodies' boats. We should try to wake up and look with the clear Eye of the water we float upon.
—RUMI[1]

Certain ideas take root in the psyches of their believers, coloring all their perceptions. Kierkegaard noticed that the less support an idea has, the more fervently it must be believed in, so that a totally preposterous idea requires absolute unflinching faith. This perverse balance helps account for the wide variety of beliefs—some "self-evident," others dogmatic—to which people attribute certainty. Abstract and profound ideas, like drawings with an unfinished quality, may contain a certain open-endedness that makes them appeal to many different people. As a virus reproduces itself by infiltrating the cell, so some notions would appear to latch onto the human imagination by being suggestive, self-contradictory, or symbolic. The great ideas leave an empty space in which believers recognize themselves. Fascinated with their own reflection, intrigued by the way a notion speaks directly to their own experience, the converted then proselytize to others on behalf of the idea and its amazing truth. Yet in reality they may be just passing a mirror and saying, "Look."

Whether they are true or not, subscription to certain philosophical notions puts hinges in the mind with which we can swing open the doors of perception. You may believe (with the Buddhist) that time, space, and individuality are illusions (perpetrated by samsara, the merry-go-round of regeneration). You may believe, as Nietzsche did, that everything you do will recur in the future an infinite number of times, or, conversely, like novelist Milan Kundera, that each act in the play of reality comes only once (floating away into "the unbearable lightness of being"). For the Nietzschean, each thought can have an immense significance: it will be repeated throughout eternity. For Nietzsche the thought of the eternal recurrence of the same raises the stakes of being, because any crisis or pain must be dealt with not just here and now, but forever. For the Kunderan, however, events and thoughts may have no special significance and may appear meaningless, arbitrary, and random, slipping into the future never to return. Because Nietzsche's idea of the eternal recurrence and Kundera's notion of the lightness of being are diametrically opposed, they cannot both be continuously entertained. Yet each dramatically colors the perception of the true believer.

Again, if you hold that your life has been preordained by God, or that interacting waves and particles whose antecedents were present at the origin of the universe determine your every thought and action, you may be inclined to act less responsibly—and more nihilistically—than if you believe you have perfect freedom of choice. Nietzsche sought to prove his doctrine of cosmic rerun with reference to thermodynamics. Using the example of a Christian belief in eternal damnation, he indicated that an idea need not be true to exert a tremendous effect. The truth or falsity is not a prerequisite for ideational power, the ability of an idea to transform a consciousness. Whether there is heaven and hell or starry void, free will or predestination, reality recurring forever or never, there will be believers. The human mind abhors uncertainty; in the absence of tutelage, whatever philosophy is current will rush in to fill its vacuum. (The disturbances generated by French philosopher Jacques Derrida's tortuous prose result precisely from his "deconstructive" ploy of making scintillating suggestions but anticipating and defusing all would-be conclusions.[2])

People ascribe certainty to their beliefs, reality to their perceptions. From an evolutionary epistemological approach, existence is hindered, discourse impeded, by the playful suspension of disbelief. So belief returns. Sheer survival requires that we arrive at and act upon conclusions, no matter how shoddily they are based. Doubt is a stranger to the human heart: to love or live we must believe—in something.

Let us explore now the perceptual implications of one powerfully riveting idea: that Earth is alive. This is one of those doors that, swung open, reveals a changed world. Many in the past have believed that the whole universe is alive. A corollary of this is that Earth's surface—our planet with its atmosphere, oceans, and lands—forms a giant global meta-organism. We can say that Earth is alive. But what does that mean?

Imagine a child of a present or future culture inculcated from childhood to believe that the planetary surface formed a real extension of his person, a child whose language implicitly reinforced this connection to such a point that to him it would seem to be not a connection but rather an equation. Such a person would make sense differently. Were nature not a dead mechanism but an immense "exoskeleton" (as the more limited exoskeleton or protective shell of a lobster is not only its house but part of its body), he would be less concerned by what we could not explain. And his perception of the organic would be altered. The arrangement of objects in his home, offhand comments by strangers, walks in the woods, cinema, and vivid dreams would all be linked to the organization of a living organism whose fullness of activity would be beyond his powers of comprehension. His ego no longer encapsulated by skin, he would experience the seas, sands, wind, and soil as numb parts of a body—just as feet and fingers, while open to tactile sensation, are yet incapable of speech and sight. The mountains between earth and air would seem to him anatomically placed, as "our" skeleton is between "our" bone marrow and flesh.

Putting ourselves in this child's shoes, landscapes, from jungles and glaciers to deserts and glens, become body parts in a new anatomy, even if, from the limited perspective of that body's minute and only partially sentient parts, the global or geoanatomy remains largely unintelligible. The incomplete sensation of the planetary surface as a live body is no more a metaphor for ignorance than the idea of a skin-encapsulated anatomy. Take an ant crossing a bare human foot. Does it perceive that it is touching a life-form? With this scale of differences, would it be able to distinguish a toenail from a rock or shell? Or, what can a bacterium living in the human gut conclude about the life-form that feeds it? Likewise, if we in our daily activities meander about upon the surfaces of a giant being, it need not be immediately apparent. Indeed, if one (not a positivist) believes in the necessity of metaphor as a system of explanation to "make known" our ignorance, then the image of a live planetary surface may itself—like Democritus's theory of atoms—be enough to launch an entire new epoch of scientific research and individual action. Though inevitably we would reach the borders upon which such a program would be based, it is possible to imagine language itself embedding the structure of such an altered state of affairs and "making it real." Blue Earth itself would color all our perceptions.

Imagine someone from this culture picnicking. She believes her environment—and not just individual plants, animals, fungi, and microbes—to be part of her self. The grass on which she sits is a patch of tissue lining the inside of the superorganism of which she forms a part. The bark at her back, the dragonflies, the birds, the clouds, the moist air, and the ants tickling her foot—all these sensations represent from her point of view not "her" but that which, from our point of view, we pedantically term the self-perception at one site of a modulated environment. Like the ants, "she" senses what is beyond "her." When "she" pulls her T-shirt over "her" knees, this is no longer human, but one locus of sensation within the kaleidoscopic entrails of a planet-sized photosynthesizing being.

The physiology is vast. The prostaglandins in people's bodies have many functions, ranging from ensuring the secretion of a protective stomach coating that prevents digestive acids from acting on the walls of the stomach to causing uterine contractions when ejaculated along with sperm in the male semen. So, too—looking at it now from an artificial position outside the physiology—the whole woman, by what she says, makes, and does, performs multiple functions within the global anatomy. A hormone is a biochemical produced in one part of the body that is transported through the circulatory system and causes biological reactions. The pituitary gland, at the base of the human brain, for example, stimulates sex hormones in the ovary and testes, causing pubic hair to grow.

As an animal breathes, it affects the entire Earth system. Water and atmosphere act as veins conveying matter and information within the geoanatomy. Indeed, the environment is so "metabolic" that minor actions may be amplified until they have major effects, while seemingly major effects may be diminished or negated. The ground is a live repository for metabolisms like the rings of tissue left in the wake of a growing tree. Treelike, Earth grows, leaving behind it archaeological and paleobiological rings. The "woman" herself is part of a currently active geological stratum; and, far from dead, the air around the body that we, from habit, distinguish as human is fluid and thriving, part of an external circulatory system exploited by life as a whole.

To one raised to believe in the textbook notion of a static geology to which biology adapts, the young woman seems to be eating alone, surrounded only by plant life. Yet from "her" perspective the environment around "her" pulses with communicative life; "she" is at a busy intersection in the heart of nature. A part of nature, "she" is not simply "human" but an action within the self-sensing system of a transhuman being. (Indeed, language's personal pronouns falsify; they do not do justice to "her" but make "us" see "her" as a "thing" in a way that is in fact alien to "her." The self-extension to the environment has altered everything.)

Bearing in mind the idiosyncrasies of "her" perception, let us return to more ordinary language on the condition that without quotations *she* and *her* are still recollected as being imprisoned in such jail-bar-like quotes by our more fractured word-biased views. With that said, there are things that to her would seem bona fide but remain quite mysterious from our worldview. Being called by a long-lost friend during the very moment she was thinking of him would not necessarily strike her as being what Carl Jung termed synchronicity—coincidences with such deep significance that one concludes they are more than mere coincidences. For her, strange coincidences come from her ignorance of the huge physiological system of which she forms a small part. Rain forests and seaside sludge she sees as vital organs, as inextricable to the biosphere as a brain or heart is to an animal; humans, however, she may construe as lucky beneficiaries of the establishment of superorganism, fluff like fur or skin that can be sloughed off without incurring major harm to the planetary entity as a whole.

Her uncle, a "geophysician," tells her that humanity has caused in the biosphere a physiological disturbance. "Earth," he says (in so many words), "is oscillating between ice ages and interglacials; it has the global equivalent of malarial chills and fevers. . . . Oil in the ground has become a gas in the atmosphere . . . tall tropical forests are being flattened into cattle . . . our vital organs are plugged with asphalt." Deserts, he tells her, are appearing like blotches on the fair face of nature. "But," he tells his niece, "we don't even know if these 'symptoms' are indicative of transformative growth—in which case we are experiencing normal 'growing pains'—or debilitating disease. Perhaps it is both, as in pregnancy, which, if encountered by a being as minute in relation to a pregnant woman as we are in relation to Earth, might be misdiagnosed as the most bloated and dangerous of tumors."

Let us adopt the mask of metaphysical realism for a moment and peer through the empty spaces, the (w)holes, which are all it has in the way of eyes. The example of Earth's physical appearance altering due to the popularity of an idea—whether true or not—is an indication of what can happen when philosophy meets technology. But all this speaking of Earth as if it had a "face" and a "fever"—as if it were some sort of comprehensible living entity—begs the question: Is Earth really alive? And if it is an "organism," what kind of organism is it? Can it think? Certainly the biosphere can be not an animal but only animal-like. And if Earth does not resemble any other organism we know, have we reason to call it an organism at all?[3]

Scientific evidence for the idea that Earth is alive abounds. The scientific formulation of the ancient idea goes by the name of the Gaia hypothesis. The Gaia hypothesis proposes that the properties of the atmosphere, sediments, and oceans are controlled by and for the biota, the sum of living beings, as discussed in chapters 18 and 19. In its most elegant and attackable form, the hypothesis lends credence to the idea that Earth—the global biota in its terrestrial environment—is a giant organism. Evidence for organismlike monitoring of the planetary environment does exist. Reactive gases coexist in the atmosphere at levels totally unpredictable from physics and chemistry alone. Marine salinity and alkalinity levels seem actively maintained. Fossil evidence of liquid water and astronomic theory combine to reveal a picture in which the global mean temperature has remained at about 22°C (room temperature) for the last three billion years; and this constancy has occurred despite an increase in solar luminosity estimated to be about 40 percent. Under the Gaia hypothesis such anomalies are explained because the planetary environment has long ago been brought under control, modulated automatically or autonomically by the global aggregate of life-forms chemically altering one another and their habitats. All these anomalies suggest that life keeps planetary house, that the "inanimate" parts of the biosphere are in fact detachable parts of the biota's wide and protean body.

On Earth, for example, temperature modulation may be accomplished, at least in part, by coccolithophores, a form of marine plankton invisible to the naked eye but shockingly apparent in satellite images of the northeastern Atlantic Ocean (see figure 19.1). These tiny beings produce carbonate skeletons as well as a gas called dimethyl sulfide. The gas, pungently redolent of the sea itself, reacts with the air to produce sulfate particles that serve as nuclei for the formation of raindrops within marine stratus clouds. The plankton, then, by growing more vigorously in warmer weather, may enhance cloud cover over major sections of the Atlantic Ocean. But the enhanced density of the clouds leads to more reflection of solar radiation back into space, so that the same plankton growing in warm weather cool the planet. In these sorts of ways the subvisible but remotely sensible beings may be part of a global system of temperature control similar to the thermoregulation of a mammalian body. Without attributing consciousness or personifying them as minute members of some global board of climate control, the organisms may be seen to act together as part of a system of thermoregulation like the one that in us stabilizes our body temperature at approximately 98.6°F. Locally acting organisms apparently can affect the entire planetary environment in a way that builds up organism-like organization.

In a way it is not so surprising that individual action leads to the appearance or, indeed, the actuality of global controls. Academically, the disinclination to accept the possibility that Earth regulates itself in the manner of a giant living being seems to have less to do with physical and chemical evidence—which lends itself to such interpretation—than it does with the status of modern evolutionary theory. Darwin considered the individual animal to be the unit of selection, but in the modern synthesis of neo-Darwinian theory, natural selection is seen as operating on genes as much as on individuals, and evolution is mathematicized as the change in frequency of genes in populations consisting of individual animals. So, too, altruism in sociobiology is often seen as the tendency of genes to preserve themselves in their own and other gene-made organisms; biologists tend to dismiss the idea that groups above the level of the individual can be selected for, because they are not cohesive enough as units to die out or differentially reproduce. Evolutionary biologists lump arguments for selection of populations of organisms with the archaic oversimplification "for the good of the species"; they then perfunctorily dismiss such arguments as misguided, if not altogether disproved. Yet, as elegant as the mathematics combining Mendelian genetics and Darwinian theory sometimes may be, sociobiologists have a deep conceptual problem on their hands with their insistence that natural selection never works at a level above genes and the individual.

First of all, it is not clear what sociobiologists think an individual is; they fail to analyze or define this term, assuming that it is self-evident because of a parochial focus on the animal kingdom. The problem is that the microscopic cells, which in the form of colonies must have given rise to the ancestors of all modern plants, animals, and fungi (as well as protoctists—algae, slime molds, protozoans, and the like), did, and still do, assume the form of individuals. How, then, can evolution not work at a level above that of the individual if the very first animals were themselves multicellular collections—populations—of once independent heterogeneous cells?

The animal body itself has evolved as a unit from a morass of individuals working simultaneously at different levels of integration. Sociobiologists and neo-Darwinian theorists disdain "group selection" because they don't have strong enough cases for its existence in populations of animals. But it may well be that, due to their large size and late appearance on the evolutionary stage, animals have not yet achieved the high level of group consolidation found in microbes. No matter how elegant the mathematics, dismissing group selection as an evolutionary mechanism requires dismissal of individual animals also, for the body of the academic itself provides a counterexample to the thesis that natural selection (if it "works" at all) never works on "groups." A person is, after all, a composite of cells.

Part of the problem here is the restrictive focus on animal evolution when animals themselves are the result of multigenome colonial evolution and represent only a special intermediary level of individuality, midway between microbes and multianimal communities. But cells, animal species, and the biosphere all evolve concurrently. The first plants and animals began as amorphous groups of cells, later evolving into discretely organized and individuated communities of interacting cells. The evolution of individual cells led to the group of cells we recognize as the animal body. Groups of animals such as insect societies and planetary human culture begin to reach superorganism-like levels of identity and organization. The human body is itself a group that has differentially reproduced compared with other, more loosely connected collections of cells. That cells of human lung tissue can be grown in the laboratory long after the victim from which they were taken has died of cancer shows that the cells in our body are tightly regimented into tissue groups but still retain the tendency for independent propagation.

To be consistent, mainstream biology should explain how something called "natural selection" cannot be "acting" on groups of organisms if the animal "individual" is in a very deep sense also a "group" of organisms, namely, cells with their proposed histories and origins. Here we can accept, for the sake of argument, that several hundred million years ago multicellular assemblages began to evolve into the animal lineage. These groups left more offspring than their free-living unicellular relatives. Their very bodies contained the principle of social altruism, in which some cells specialized and curtailed their "selfish" tendency toward indefinite propagation for the "benefit" of the group to which they belong.

Working within the framework of evolutionary theory, we must accept the argument that "group selection" exists in the origin of animals.

Therefore, we must (again, within this framework) concede that evolution favors populations of individuals that act together to re-create individuality at ever higher levels. This somewhat freaky assertion calls into question the very usefulness of trying to isolate the units of natural selection: because of the articulation or community relations of living things, the differential reproduction of units at one level translates into the differential reproduction of units at a higher, more inclusive level. I anticipate that the mathematical theory of fractals, in which the same features are present in interlocking geometrical figures at various scales of analysis, may be useful in illustrating the principle of emergent identity in the series cell, multicellular organism, and superorganismic society. In principle, the "animal-like" nature of Earth can be considered fractally as resulting from the Malthusian dynamics of cells reproducing within a limited space.

If this chapter's evolutionary understanding (qualified by placement under the rubric "metaphysical realism") is "right," it may be that Earth itself represents the most dramatic example of emergent identity. The properties of global regulation on Earth result from the metabolic activities of the organisms that comprise our biosphere as in Lovelock's Daisyworld (see chapter 18). On a less inclusive scale, "global" human consciousness and unconscious physiological control mechanisms can be traced to the synergistic effects of millions of former microbes acting locally to comprise the human body and its central nervous system. As an individual, the human body has evolved in isolation from other organisms, whereas the biosphere as a whole does not even have as clear a physical boundary separating it from the abiological cosmic environment, let alone from other organisms. In this sense, the biosphere is much less an individual than is any animal. But the lack of biospheric individuality may be as artifactual as it is temporary. A superorganism as large as Earth has not had the chance to evolve distinctive characters in isolation. Moreover, even if it were far more complex (anatomically, physiologically, and "psychologically") than a mammal, we may have difficulty understanding it precisely because of that complexity. Because Earth is so huge, the Gaian organism may not be as apparent—or as consolidated—as a single animal. Over time, however, the Gaian superorganism can be expected to consolidate and become increasingly apparent; it may in the next centuries even become "obvious" to the majority of human beings.

Russell L. Schweickart, a NASA astronaut from 1963 to 1979, was an adviser on Biosphere II, a private-capital project to build a multimillion-cubic-foot biosphere near Tucson, Arizona, for about the price of a modern skyscraper. "The grand concept," he said at a 1980s meeting of those working on the project, "of birth from planet Earth into the cosmos—when, 1993, 1994, 2010, 2050, whenever—is a calling of the highest order. I want to pay a lot of respect to everyone associated with that grand vision for their courage to move ahead with this in the face of the unknowns which make the lunar landing look like a child's play toy. There were a lot of complexities there, but we were dealing with resistors, transistors, and optical systems which were very well understood. Now we're wrestling with the real question: that natural process of reproduction of this grand organism called Gaia. And that's what all the practice has been about."[4] Many astronauts spacewalking or gazing at Earth report on the tremendous transformative power of the experience. That looking at Earth from space could so totally change a person's consciousness suggests that the experience has not yet fully registered upon the body politic. People such as Schweickart who have seen Earth from "outside" in space may be more prepared to accept the unorthodox idea that the biosphere not only is a living entity but also is about to reproduce, as many individuals—and, indeed, many cellular groups arranged into individuals—have done "before."[5]

However, at the Cathedral of St. John the Divine in New York City in 1987, the thoughtful plant geneticist Wes Jackson protested the idea of Gaia on the basis of Gaia's "infertility." Jackson claimed there is no way Earth could be an organism because all known organisms, from microscopic amoebae to whales, reproduce. Because Earth has no "kids," it cannot be a real organism. It is only a metaphor, he said—and it may even be a bad one. According to Jackson, we do not even know what Earth is. ("What is God?" he asked provocatively, suggesting the questions were similar.)

In a way I do agree with Jackson. Earth seems indefatigable in its capacity to make us wonder about its true nature. Yet I had become convinced that Earth is, in a sense, reproducing before ever hearing Jackson raise this counter-Gaian argument. The reason for my conviction that the biosphere is on the verge of reproduction has to do with two things: (1) The growing number of scientists and engineers involved in designing, for a variety of reasons, closed or self-sufficient ecosystems in which people or aggregates of life can live; and (2) my assumption that humanity is not special but part of nature. For if we are part of Earth, so is our technology, and it is through technology that controlled environments bearing plants, human beings, animals, and microbes will soon be built in preparation for space travel and colonization. In space these dwellings will have to be sealed in glass and metal or other materials so that life will be protected inside them. Such material isolation gives the recycling systems discrete physical boundaries—one of the best indications of true biological "individuality." Thus, the bordered living assemblages necessary for long-term space travel and planetary settlement by their very nature bear a resemblance to biological individuals at a new, higher scale of analysis. They look startlingly like tiny immature "Earths"—the biospheric offspring Jackson claims must exist for Earth to be a true organism.

We can trace a progression in size in these human-made containers of recycling life. Clair Folsome, University of Hawaii, kept communities of bacteria and algae illuminated but enclosed in glass. They stayed healthy and productive from 1967 until the mid-1980's. There is no reason to think they may not be immortal despite being materially isolated from the global ecosystem. Similarly, Joseph Hansen of NASA developed a series of experimental desktop biospheres consisting of several shrimp, algae, and other organisms in sealed orbs half filled with marine water. These last for years, and in some crystal balls the hardy shrimp inside even reproduced. On a still larger scale, private and governmental space administrations in the Soviet Union, the United States, Japan, and other countries are developing the art of creating materially closed perpetually recycling ecosystems. Crucial not only to space travel and colonization, these miniaturized ecosystems could also protect endangered species, maintaining air, water, and food supplies, and allow, in the long term, the possibility of social, cultural, and biological quasi-independence on the ever more crowded and homogenized Earth.[4]

If successful, controlled ecosystems will carry a powerful educational message about the need for cooperation of people with one another as well as with the other species that support the global habitat. And if perpetually recycling ecosystems can be erected and maintained, a whole new scientific discipline may arise from the possibility, for the first time ever, of comparing "parent" and "offspring" biospheres. Former astronaut and physicist Joseph Allen points out that the quantum-mechanical revolution that so marks modern physics derives from the comparison by Niels Bohr of helium and hydrogen nuclei; having more than a single biosphere to observe may likewise revolutionize biology.

Communication established between two semiautonomous biospheres may resemble in emotional impact the relationship of a mother or father to a daughter or son. Yet the "children" will teach: the safe modeling of potential ecological disasters within a new biosphere may provide dramatic warnings and even perhaps usable information on how to ward off the environmental catastrophes—from acid rain to pesticide contamination of foods—that potentially await us. New biospheres thus may serve as living whole-Earth laboratories or "control worlds," inaugurating differential reproduction on the largest scale yet.

The importance of the development within the biosphere of such enclosed ecosystems cannot be overestimated. Whether or not individual, national, or private venture capital models succeed or fail is irrelevant. What we see, rather, is the tendency of Earth (or Gaia, or the biosphere) to re-create itself in miniature. Because we, from an evolutionary perspective, are natural and not supernatural creatures, Earth is, through the high-tech expedient of modern world civilization, re-creating versions of the global ecosystem on a smaller scale. To some the view of Earth biospherically splintered into semiautonomous ecosystems would be a technocratic blunder equivalent to the formation of a planetary Disneyland. But even if Earth is saved as a single biosphere, such materially closed ecosystem technology will be necessary for extended human voyages into space or the settlement of off-world sites for emigration or long-term exploration. Thus, we do seem to be caught in precisely that historical moment when Earth is begetting its first, tentative batch of offspring. That humankind is currently the only tenable midwife for Gaian reproductive expansion is a gauge of our possible evolutionary longevity and importance—provided that the violently phallic technology that promises to carry life starward does not destroy its makers first.

The Gaia hypothesis is at once revolutionary science and an ancient worldview, with the power to spur not only scientific research but religious debate. If we take it to its logical extremes, it says not only that Earth is alive but also that it is on the verge of producing offspring. From a strict neo-Darwinian perspective, this may be a mystery, for how can a giant organism suddenly appear ex nihilo and then just start reproducing? Yet from a broader philosophical perspective the reproduction of the biosphere makes perfect sense. We are animals whose reproduction is an elaboration of the reproductive efforts of cells; the organismic and reproductive antics of Earth have not appeared in an evolutionary vacuum. Gaia's weak, immature attempts at "seed" formation and reproduction result from the sheer numbers of organisms reproducing at Earth's surface. What before occurred in the living microcosm of cells is now transpiring in the larger world of animal communities. The Malthusian tendency to increase exponentially in a limited space beyond the resource base apparently may account for more than just the evolution of new species; it leads also to the appearance of individuality at ever greater levels and scales of analysis.

This essay broaches what might be termed a Nietzschean ecology. That is, it attempts to hint at an art of biology whose unveiling may be as important as biology itself, at least in terms of biological understanding as it applies to the "individual" in his, her, or its restless search for meaning. (Academicians, guard your territory!) The appearance of closed "offspring" biospheres from the original open biosphere repeats or continues the process by which "individual" plants, fungi, and animals appeared from communities of microbes. As the folk saying goes, *Plus ça change, plus c'est la même chose*: the more things change, the more they stay the same. As Nietzsche scrawled in one of his notebooks: "Everything becomes and recurs—forever!"

As we have seen, even a false idea may color our views of the world, and where there is a chance of changing the world, there is the chance of bettering it. Gaia is such an idea, yet one with the added punch that it may be proved true. (Oscar Wilde observed that "even true things may be proved.") It was interesting to watch the debate in March 1988 as the American Geophysical Union (AGU) met in San Diego to "test" for the first time among polite scientific society the validity of Lovelock's hypothesis. In fact, as everyone saw in the epistemology session (and any sort of philosophical discussion is rare at scientific meetings these days), it was fairly easy to show that Gaia is not, strictly speaking, testable. Whether one took him to be a very naive epistemologist or an extremely sophisticated sophist, James W. Kirchner was correct when he compared the postulate that Earth is alive to Hamlet's theory that "all the world is a stage." There is no way of proving or disproving such general notions. Kirchner pointed out that Gaia is not a valid hypothesis because it does not say something we can verify or falsify, something such as (Kirchner's example), "There are footlights at the edge of the world."

In fact, Gaia is not a hypothesis. It is, like evolution, a metaphysical research program. The idea that Earth is alive is extremely fruitful, able to suggest many scientific models and lines of inquiry. Yet ultimately it is unprovable, a matter, at bottom, of faith. It is, after all, a worldview. What positivists miss in their attack on Gaia is that they, too, are up to their necks in metaphor and metaphysics. There is no avoiding metaphor and metaphysics. When worldviews collide, weak ones are obliterated in the encounter. In my view, what happened at this conference was an encounter of worldviews. But it was no head-on collision. Rather, the old panbiotic or animistic worldview (at the center of the Gaia hypothesis) sneaked its way into mainstream discussion. In a direct confrontation, the Gaian worldview would have been eaten alive by the prevailing world-view (atomistic science and its Platonic "laws" as absolute reality). But by disguising itself as a testable hypothesis, Gaia was smuggled into a presti-gious scientific discussion. We would never expect the discussants at a serious scientific conference to bring up as the main question their own view of reality. But this is, in effect, what happened. Like the soldiers in the Trojan horse, the Gaian worldview sneaked past the well-armed guards of metaphysical realism ("science") by disguising itself as a hypothesis. And now the worldview Gaia, having lodged itself inside the worldview metaphysical realism, is impossible to extract without damage to both. Our entire conception of life and its environment is being called into question. What is life? Technology? The environment?

Perhaps another Greek myth, because it has not strayed onto the dan-gerous battlefield of truth, better sums up the present philosophical situ-ation: Once Narcissus stood and eyed the still waves that reflected his own image. He had never seen himself before. He became infatuated. And now we gaze in the looking glass of satellite imaging technology. Again we see the water. Again . . . but what is "ourselves"? And who—or what—is *this body*?

Chapter 21 Notes

1. The quotation is from Rumi's *We Are Three* (translated by Coleman Barks). Jalal ad-Din ar-Rumi Rumi (1207–1273) was a Sufi love mystic who wildly spun around as he delivered his musical verses, which were transcribed by assistants. He was the first "whirling Dervish," and it is claimed that his poetry read aloud in the Persian original is so musical it sends listeners into a trance by its aural quality alone.

2. The technique of leading people in certain directions and then "pulling the rug out from under them" resembles the method of the sleight-of-hand artist. Both the deconstructionist and the magician present signs that are typically organ-ized or mentally ordered into a narrative of events. A difference is that, whereas the exponent of legerdemain presents approximately the minimal number of sensory stimuli to arrive prematurely and mistakenly at a certain impression of reality, and this impression is then revealed to be "wrong" (that is, clearly only an image) after the performance of the "trick," the deconstruc-tionist uses language as the medium for the presentation of mirages that are

more or less continuous; the deconstructionist does not entertain like the magician with a series of discrete and contained surprises but reveals that the attribution of "finished" images and mirages from unfinished signs and stimuli proceeds unceasingly. The difficulty with deconstruction is that it shows off-stage, whereas traditional magic shows onstage. But this difficulty has to do with the "broadening" of the stage, the spilling over of theater into the realms of everyday life: It cannot be gotten rid of by dismissing as unreadable all deconstructive prose. Clearly, the conclusions arrived at through the use of language, and especially of "language with ordinary words," may be as bogus as the conclusions arrived at through the motions of a sleight-of-hand artist— and especially one manipulating not apparatus onstage (where the theatrical element is expected) but small, ordinary objects such as cards and coins in the home space so normally above suspicion.

3. We say all this keeping in mind that our language—and our science—bears within it its own deeply embedded and usually unexamined set of metaphysical assumptions. Derrida has unequivocally shown this. Just as Nietzsche did not need thermodynamics to be affected by the idea of eternal recurrence, one need not justify the culturally marginal notion of living Earth by reference to or with the sanction of a cultural mainstream, a tradition of knowledge not at home with such ideas. Nonetheless, the possibility of scientific sanction indicates the reality of the approach of this notion into the mainstream.

4. Sagan, D., 1990a.

5. Part of the problem with the whole concept of evolution—and all narrative "explanations"—may be the unexamined reliance upon the unprovable assumption of linear time, a logocentric assumption. The verb tenses of languages perpetuate the assumption of temporality. The relation of language to the bias of linear time is here dubbed "chronic." In fact, the relationship of life-forms may be better seen as four- or multidimensional, in which case the evolutionary unfolding in linear time is better seen as only a "slice" through true space time.

The Pleasures of Change

DORION SAGAN AND ERIC D. SCHNEIDER

Larger than symbiogenesis (Part II: Chimera) and the evolutionary history of life is the idea that life is only one of several natural thermodynamic processes. All complex systems, in the real not the virtual world, require continuous flow of energy. All reduce gradients, or differences across a distance. "Nature abhors a gradient" is Schneider's extension of the second law of thermodynamics to apply to open systems. A tornado reduces an atmospheric pressure gradient; life reduces an electromagnetic solar gradient. Both cycle matter and, for a time, grow. Both have a form that is also their function—to reduce gradients. Life and tornados don't just resemble each other. Rather they belong to the same class of naturally complex thermodynamic systems.

The strange fact that the human mind is able to imagine eternity—geometric shapes, numbers, and other changeless, self-identical forms—has had a dramatic impact on how we think of reality. In ancient Greece, Pythagoras led a cult that worshipped the triangle, a perfect form, and believed that numbers had a separate existence outside of time. Later Plato was enthralled with the notion of incorruptible Ideas beyond this world of aging and death, of rust and dust and ceaseless change. Through Aristotle the notion of a Platonic realm of timeless perfection influenced the Church, informing the traditional Judeo-Christian notion of Heaven, which, as David Byrne sings, "is a place where nothing ever happens." But how curious it remains that we can imagine the timeless, as epitomized by the mathematical equations of the physicist who has tried, ever since Newton, to get into the mind of God to discover the eternal laws behind this world of change. Curious because we live in a world of change where perhaps the most truly uncorruptible, eternal, and changeless thing is our ability to imagine such stability! Plato hypostatized such thought into a higher, Heaven-like plane of existence, of eternal, unchanging ideas of which this real world of changing things is only an imperfect, corruptible copy. In fact, despite the undisputed usefuleness of the eternal laws of nature, science is now coming round to a more realistic view of a thoroughly evolutionary universe, permeated with change, in which even the laws of the universe may alter. In this universe the second law of thermodynamics, which describes irreversible processes in nature, has a special significance.

The Greek philosopher-scientists were deeply influenced by Parmenides, who preached Being, and who had a greater influence on the course of Western thought than did his intellectual rival Herakleitos, the poet of Becoming. But in a way we are now coming round to Herakleitos's way of thinking. Herakleitos taught the importance of change and liked to point out that even things that seemed the same were different later in time. "You cannot step into the same river twice," said Herakleitos in his most famous fragment. Herakleitos was referring to the particles of water, replaced and different each time, though the river may seem the same. And each of us is such an everchanging river: every five days your stomach grows a new lining, a new liver comes every two months, and the biggest organ of your body, your skin, is itself entirely replaced every six weeks. What we see as identity is really an illusion. "You" are unintermittent biochemical turnover, a ceaseless swirling of organic change.

Today we live in an evolutionary universe, where change is endemic to how we see the universe and ourselves. In this essay we will make the case that the sort of change we call biological evolution is in fact a natural outgrowth of a more fundamental tendency toward change belonging to the inanimate universe. This is the tendency—described by the second law of thermodynamics—for gradients to break down, differences to dissolve, and things to become more disordered over time. A drop of cream in your coffee will spread out, never to return to its original state. There are many more ways for the cream molecules to be mixed with the coffee molecules than for them to be separate. Ever since Ludwig Boltzmann (1844–1906) wrote the equation linking change to probability that is engraved upon his tomb, modern thermodynamics—from the Greek *thermos*, meaning "heat," and *dynamos*, meaning "movement"—has based its understanding of how things become more mixed up over time on probability theory: energy does not localize but spreads out, fine differences left alone do not become finer but tend to degrade into uniform states, and randomized confusion is far more likely to occur on its own than states of increasing differentiation, elegance, organization, and apparent design. A living being or a work of Shakespeare would never be expected on the basis of chance alone. Yet such complexity exists. Somehow change also moves in a direction of increasing complexity, as in the growth of a red maple tree from a spinning seedling, the succession of organisms seen in ecosystem development, or the finished production of this glossy book. At first glance such things shouldn't happen, as the second law mandates that they should be doing the opposite, falling apart instead of evolving into the more and more complex processes seen in nature. We name the enigma of evolutionary complexity in a thermodynamic universe the "Schrödinger paradox" after Austrian physicist Erwin Schrödinger, who first brought it to public attention in a 1943 lecture series at Dublin's Trinity College. Time has a forward direction, paralleling the irreversibility of the second law, originally put forth as a formalization of engineers' observations that steam engines were inevitably inefficient, losing work inescapably to heat. Time does not go backward, the three-dimensional equivalent of a film projected backward, or we might see windows unbreaking and made whole, eggs being unfried and slurping up against gravity, reassembling with reverse jigsaw-puzzle precision into the shells in the cook's hands. We don't find find smoke gathering into the end of your father's lit cigarette, or your life would poetically culminate with orgasmic pleasure rather than painful death.

It turns out that the appearance of local complexity and organization, as epitomized by living beings, is itself a way of aiding randomization, abetting the second law, the uniform state "craved for" by nature. A tornado is a highly organized natural structure, and its "purpose" is to reduce the barometric gradient between high and low pressure air masses—and the tornado will spontaneously organize to accomplish that purpose. Once the air masses reach their more probabilistic state of calm, the complex whirling tornado has fulfilled its function and disappears. Startlingly, life itself also so much displays such traits of fulfilling natural unconscious purposes that one is led to wonder whether our conscious and unconscious drives and desires do not ultimately reflect a thermodynamic imperative. As thermodynamicist Jeffrey Wicken says, "I see no way to dodge the Kantian challenge. In Kant's conception an organism was a 'natural purpose' in which each part and process was jointly cause and effect, ends and means, of the operation of the whole." Just as nature abhors a vacuum, she dislikes differences and will destroy them over time, achieving end states of higher probability and more likely distribution of components. Our very biological urges to eat, drink, and mate—related to reproduction and growth—have their roots in the inanimate world that often grows complex structures locally to break down larger-scale anomalies. It is the temporal nature of thermodynamic processes to have an end in equilibrium, a state of stasis, a point of no further changes in state, which gives all natural systems a direction, an end point. And it is in the achieving of this equilibrium end point that nature crafts, from available materials if conditions are right, equilibrium-creating systems. Such is the tornado, a complex cycling storm that equalizes the pressures between conflicting air masses.

Living beings are complex equalizers of another kind. We thrive on the sun, necessarily excreting liquid, gas, and solid waste while giving off low-grade heat, as is our thermodynamic wont. So, too, our technologies run on high-quality fuel that becomes low-quality exhaust and pollution. The improbability of being you is high. But you are as organized as you are because you are part of a planetary thermodynamic system—the biosphere—that has been continuosly trapping and rerouting the high-quality energy of sunlight, and producing the low-quality energy of heat, for over three billion years. The replicative machinery of the DNA molecule does not exist in isolation but ensures the transgenerational stability of energy-degrading systems. If life could, it would take all the energy of the sun and turn it into itself. From a cosmic perspective, the changes we see in evolving life reflect nature's tendency to destroy differences—specifically the difference between the energy-rich sun and the energy-poor space around it. Life helps spread the sun and would—like the tornado without the differentiated air masses around it—cease to exist in the sun's absence. Life openly processes energy from its surroundings. But why and how do such complex things as tornados, puppies, and trees emerge from this sameness of the equilibrium world of thermodynamics?

Just as Charles Darwin showed the kinship of humans to other life-forms—startling Victorians with the notion that people were not specially created but evolved gradually from "lower" animals—so today we realize the kinship of biological change to change in nonliving matter: people are not only evolving animals, but the growth and evolution of living matter obey rules of change that apply also to nonliving systems. Thermodynamics, which is the study of energy balances and flows in simple and complex systems, governs many of the important changes we see in living systems. Life, the universe, and everything are not static but charged through and through with a mandate to change implicit in the second law of thermodynamics. Cultural critic C. P. Snow said that not knowing the second law is like not having "read a work of Shakespeare's."[1] English physicist Arthur Eddington wrote, "The law that entropy increases—the Second Law of Thermodynamics—holds, I think, the supreme position among the laws of Nature. If someone points out to you that your pet theory of the universe is in disagreement with Maxwell's equations—then so much the worse for Maxwell's equations. If it is found to be contradicted by observation—well, these experimentalists do bungle things sometimes. But if the theory is found to be against the Second Law of Thermodynamics, I can give you no hope: there is nothing for it but to collapse in the deepest humiliation."

In layman's language the second law basically says that energy inevitably degrades; that the high-quality fuel and energy available for work will inevitably be lost unless it is replaced; that pollution—and death, decay, and attrition—are inescapable. (Death is simply what happens when the stuff of life—fostering equilibrium with the environment—itself comes to equilibrium.) A more technical definition is that entropy (a mathematical measure of such disorder, originally defined as heat divided by temperature) tends inevitably to increase in isolated systems. Unlike us, who totally depend upon a flow-through of matter and energy (food, water, air), isolated systems are sealed off to matter and energy exchange with the outside world.

Today we find that thermodynamics has a lot to say about life. But on the surface thermodynamics would seem to contradict life. Because thermodynamics was formulated during the nineteenth century, when railroad and ship steam engines determined the superiority of industries and navies, its original application was to machines, which are substantially different from living beings. French physicist Nicolas Leonard Sadi Carnot (1796–1832), the estranged son of a Napoleonic minister of war, was irked that England produced more-efficient steam engines than France and initiated the first studies of thermodynamics. Before his notebooks were burned by order of the state to decrease chances of contagion by his infectuous cholera, Carnot scribed that it was not simply the temperature of the steam-producing boiler that made pistons pump hard and fast, but rather the difference between the temperatures of its boiler and that of its radiator. "The production of heat is not sufficient to give birth to the impelling power," Carnot wrote in his booklet *Reflections on the Motive Power of Fire* (1890; 2005), "it is necessary that there should be cold; without it, the heat would be useless."[2] Heat must *flow* from hot to cold, and it is the gradient, the difference of temperatures over a distance, that sets up the conditions for the flow to take place. By making engines with greater gradients, industrialists increased the efficiency of their heat engines.

In contrast to engines, living beings do not turn on and off but continuously process energy to make themselves; they take in liquids, solids, and gases from the outside not only to function but to maintain and reproduce their organization. Biologists' technical term for such behavior is *autopoiesis*, from the Greek words "self" and "making" (*poiein*, as in poetry); as an automobile moves itself, so a living being metabolically makes itself. Living beings are natural purposes, "designed" by nature. In part because of their animate and mechanical subject matter, in part because thermodynamics was developed at almost precisely the same time that Charles Darwin popularized the theory of evolution by natural selection (in the mid nineteenth century), but mostly because of their distinct prognostications with regard to time, these two sciences of change—evolutionary biology and thermodynamics—came to be seen as opposites.

Evolution inherited from religion the notion of a ladder of being, ascending from lower organisms upward and outward to the angelic heights aspired to by man, a divine beyond which in evolution was replaced by the secular prospect of man's infinite future perfectibility. Space—beasts below, deities above—became mapped onto time. Thermodynamics, whose picture of the future was taken from a gritty look into the inner clunkings of imperfect steam engines, had a less starry-eyed view. Its bleak prospect included an icy Armegeddon, a universal end of cold and unrecoverable loss; one popular book from the early days of thermodynamics depicted old man Death with icicles in his beard, staring out wanly at an ocean frozen in mid-wave.

Considering that machines undergo wear and tear, break down, and ultimately attain a state of complete disrepair, how is it that life becomes more complex over millions of years of evolution? The apparent contradiction between complexifying life and inanimate processes naturally, and sometimes literally, running out of steam is a favorite rallying point of creationists. They argue that life's complexity can be explained only by intelligent design. Even former Vice President Al Gore asked biologist and complexity theorist Stuart Kaufmann at a meeting on the future of national science education how life can become more complex in a universe tending toward equilibrium and disrepair. Kaufmann had no ready answer.

THE SCHRÖDINGER PARADOX

The answer to the apparent paradox was first recognized by Erwin Schrödinger, the same Austrian physicist who won the Nobel Prize for his work on quantum physics and who invented the famous thought experiment of a cat who is neither dead nor alive. In a series of 1943 lectures at Trinity College in Dublin, Schrödinger argued that life is not some mysterious stuff with special properties somehow beyond the normal operations of chemistry and physics.[3] In this set of three lectures, later printed as the classic *What Is Life?*, Schrödinger combined chemistry and physics to put biology on a different track. He saw two processes in life: order from order, or genetics, and order from disorder, or thermodynamics. At the center of life he saw the gene, with its yet-to-be-discovered DNA that produces order from order, progeny as an almost perfect replicant of the parent. His prediction that life must be predicated on some chemical substance helped spur the later discovery of the DNA double helix and its replicating chemistry. Noting that in a thermodynamic world tending toward decay one would not expect the highly improbable structures of living beings, Schrödinger postulated that the kind of organization, order, and complexity we see in life must somehow be imported from outside living beings. Schrödinger argued that living beings resist falling into disrepair by feeding on the high-quality energy of the sun. Temperature maps of Earth taken from space satellites have confirmed Schrödinger's general view. The most complex, biodiverse, and "organized" ecosystems—the rain forests of the Amazon—are the best air conditioners on the planet. The trees that are at the heart of these ecosystems process massive quantities of sunlight, radiating heat to space primarily by producing water on their leaves. The water evaporates, making clouds that keeps the rain forests cool while generating entropy as heat to space. Thus, although entropy rises in isolated systems, living beings are the antithesis of isolated systems: they are open systems, using sunlight or matter made from sunlight (food) to generate entropy and heat as waste. The marvelous complexity of the jungle on Earth is made possible by life's reducing the solar gradient and sending heat off into space.

EATING THE SUN

As Schrödinger emphasized, since life is not exempt from the laws of physics and chemistry, its fascinating complexity, and the maintenance thereof, must have a source. This source, of course, is the sun. Life's basic activity is to convert low-entropy, short-wavelength photons of visible and ultraviolet radiation and reradiate them as longer-wavelength infrared radiation. The thermodynamic process of the biosphere is to turn solar radiation into living matter and heat.

An apparent contradiction to this might seem to be those beings, and indeed, whole ecosystems, that live in the absence of direct sunlight. Such creatures of the night include translucent cave dwellers with permanently sealed eyes, blind catfish holed up in Texan water supplies, and bacteria thriving in boiling hot springs in Yellowstone Park. There are also the recently discovered bacteria living thousands of meters beneath the surface of Earth inside solid rock, as well as the pogonophorans, red medusoid worms of the deep sea that undulate about in dark gardens where magma and gas from Earth's interior effervesce through cracks in the ocean bottom. But these beings also tap into gradients, albeit chemical ones (either of food or naturally occurring chemical reactions) producing entropy as they reduce differences in nature. Although they may not require direct sunlight, they do require energetic gradients in order to maintain their own metabolism, reproduce, and evolve.

The great majority of beings in the biosphere require sunlight or the beings that trap sunlight, converting it into plant bodies, or the bodies of organisms that eat plants: in the main, the biosphere remains a solar phenomenon, tied to light. The most important chemical reaction to develop in the history of life was photosynthesis. In photosynthesis, photons from the sun are hijacked, their energy co-opted to run life and power the building of the biochemical stuff we call bodies. Today, virtually all surface life, not just plants, is dependent on this photosynthetic hijacking of photons. Animals like ourselves eat either the photon-made plants or the animals that eat them. Fungi, which are not plants but grow by absorbing organic compounds such as rotting wood or keratin, an animal protein in hair and skin, also ultimately feed off those who feed on the sun.

And this feeding off the sun—which connects all life in intricate networks of energy exchange as high-quality energy photons are sent, like wayward angels, through a maze of earthly tasks—is a preeminently thermodynamic process. Life, far from violating the second law, exemplifies it. Sucking water through their roots to their leaves, where it evaporates, the "sun-eating" trees of rain forests are at the center of a process that generates coolness and complexity locally while producing heat waste farther out. Thermodynamically, this production of heat waste is not only the paradigmatic operation of the second law but also the paradigmatic operation of life. Life is cool, you might say, and life likes it cool: but it does so because local coolness, in a thermodynamic universe, entails neighboring heat a little farther out. Temperature interpolations of Earth between Mars and Venus suggest our planet is much cooler than it should be. Carbon dioxide, the famous greenhouse gas, accounts for the majority of the atmospheres of Mars and Venus. On Earth, however, it has been systematically sucked from the atmosphere by life, which uses carbon to produce the carbon-hydrogen compounds of organic bodies. The result has been planetary air conditioning—the lowering of the global mean temperature of the biosphere—which, of course, entails the heating of the area around the Earth. Even more important to planetary cooling is the evapotranspiration of trees, producing light-reflecting clouds over them. Spreading heat, life helps reduce the solar gradient. Life is a solar thermodynamic nexus whose evolution is a transmutation of the radiation of the sun. Evolutionary complexity and the thermodynamic degradation of differences are thus not incomprehensible and polar opposites. They are, rather, part of a single process of cosmic change.

THE INEVITABILITY OF CHANGE

In his work early in the twentieth century, the Russian scientist Vladimir Vernadsky studiously avoided the term *life*, which he considered distracting, speaking instead of living matter, which forces us to focus on physical and chemical process rather than any preconceived notion of a special kind of matter.[4] Life is a kind of star stuff; it is, as Vernadsky emphasized, a solar phenomenon, a kind of "green fire," and we should properly conceive the biosphere not only as a planetary system but also as an Earth-solar system. We are, as Vernadsky liked to say, "children of the Sun"—part not only of the biosphere but of the Earth-solar system. Life as we find it on Earth is an open system connected through Vernadskian space and Darwinian time to cosmic radiation and the earliest life-forms. Among these were early photosynthetic cells, purple bacteria with metabolisms distinct from those of the green bacteria, algae, and plants that have since inherited much of the real estate of Earth's surface. Comparative analyses suggest that these ancient purple bacteria, like their cousins today, used hydrogen sulfide (H_2S) rather than water (H_2O) to make their organic (carbon-hydrogen) bodies. The hydrogen sulfide, spewed from volcanoes, would have been more plentiful on the early, more tectonically active Earth. But as Earth cooled and tectonic activity subsided, sulfide was less available for the bacteria that employed sunlight to metabolize it into their own bodies. Mutations appear to have allowed some bacteria to alter the chemical reactions of their metabolism such that they could now, using the energy of sunlight, break down the hydrogen bond in water to make their bodies. Life had been living in water since its origins three to four billion years ago. But the use of water as a metabolic resource led necessarily to an entropic waste: oxygen. Powered by sunlight, early life discarded oxygen, which reacts strongly with the carbon-hydrogen compounds of living bodies, into the atmosphere. The rock record of rust (oxidized iron) and uranium oxides demonstrates that significant quantities of oxygen did not begin to build up in our atmosphere until about two billion years ago. Today, of course, you cannot read, use your finger muscles to turn the page, or think without the metabolic energy of oxygen gas. Today Earth's oxygen-rich atmosphere, combined with the hydrogen-rich (chemists say "reduced") compounds of life, has energized the entire surface of Earth, making it something like a flashlight battery, which generates electricity because of just such difference between its positive and negative poles. The solar gradient has been converted into "redox potential"—the energetic gradient between acid and baseness, which sets up conditions for an energizing flow of electrons, as in lightning or, closer to home, the chemoelectric activity of your firing neurons. In evolutionary retrospect we can see that the oxygen-energized atmosphere of Earth—so different from the "reacted-out," mostly carbon dioxide atmospheres of our planetary neighbors Mars and Venus—is a result of living chemistry, itself an entropy-producing transmutation of sunlight.

Life's difference from the universe around it, and its tendency to evolve toward higher levels of complexity, reflects its status as an open, entropy-generating system. Using energy and inevitably producing waste as a result, life changes to metabolically stay the same, as well as to reproduce, which is itself an extension of metabolism. But life also changes because of the toxic effects its growth has on itself. The cyanobacteria that first mutated to use the hydrogen in water for their foodstuff were also the first to be poisoned by its toxicity. The inevitable production of wastes by open thermodynamic systems means that living matter cannot rest or settle like a slacker but must always be open to new challenges of change; life must be looking one step ahead of itself or suffer the consequences.

Change, both in the sense of its difference from its surroundings and in the sense that it must quickly adapt to the inevitable results of its own entropic wastes, is intrinsic to life.

THE PLEASURES OF CHANGE

Living matter belongs to a class of what Nobel laureate Ilya Prigogine, a Russian-Belgian chemist, calls "dissipative structures."[5] Such structures funnel energy through themselves as they grow and change. The first selves were likely naturally occurring chemical reactions but separate from their surroundings by a bilipid layer, a amphiphilic membrane that let in some materials and produced others as it continued to function. As thermodynamicist Harold Morowitz points out, such a membrane not only occurs naturally but also keeps oily living matter on the inside and water on the outside, and so is the perfect arena for the origins of life.[6] Dissipative structures dissipate; they are not things but processes. Thus the pleasures and challenges of life, both personally and cosmically, are not in achieving some sort of final stasis, some steady-state heaven or nirvana of eternally solved problems, but in dealing with energy flow and the change it inevitably entails.

Paradoxically, however, this continuous turnover process leads to stability; organisms change to stay the same because by doing so they maintain their gradient reduction capabilities. Early in the evolution of the solar system, for example, Earth and the inner planets were surrounded by hydrogen gas, the same element that condensed to become the sun. But hydrogen, the lightest element in the periodic table, escaped from the inner solar system after the sun's momentous thermonuclear ignition—after it "turned on." Increasingly rare around the inner, not-so-massive planets, hydrogen gas stopped only when it reached the outer planets, with their massive gravitational pulls. Some bacteria today—testifying to Earth's environment three billion years ago—can still use free hydrogen gas to make their bodies. But they are holdovers. Early on hydrogen became scarce, presumably creating one of the first of many metabolic crises for our microbial ancestors. The descendants of these beings, as we have seen, were (again, presumably) the early photosynthetic bacteria who turned to hydrogen sulfide in the absence of hydrogen, producing sulfur as their waste. Then, in the aftermath of the subsiding of tectonic activity and volcanically produced hydrogen sulfide, water-using cyanobacteria evolved, producing the modern atmosphere. But these beings, releasing oxygen, poisoned the anaerobic beings, the only sort of life alive up to then. Life poisons itself and, in carnivory, breaks down its own gradients. The resolution of each new metabolic crisis brings about another crisis, which must in turn be resolved, ultimately creating conditions under which advantages accrue to those with creativity, perception, and intelligence.

The production of volatile oxygen, at first universally deadly and to this day fatal to anaerobes such as the bacteria that dwell within cow rumens, in rocks, and beneath the muds, created yet another crisis and challenge for evolving life. Today, oxygen-breathing animals such as ourselves not only tolerate the oxidizing atmosphere but also depend upon it for our thought and movement. Tiny, matrilineally descended inclusions in our cells called mitochondria are the actual oxygen metabolizers. Genetic evidence suggests with near incontravertible accuracy that these mitochondria—found in the cells of all fungi, plants, and animals—have descended from lineages of respiring bacteria that were among the first to thrive in the blue, oxygen-rich atmosphere created by green bacteria. Today, we humans—the most spectacularly successful large mammals in the history of Earth—have created pollution crises by our own spread, and by the spread of our technology. But like the bacteria and other life-forms before us, the changes we wreak upon the global environment are the result of our capture and use of energy, which inevitably creates wastes, in accord with the second law. Trapping and rerouting the sun's photons—and even using solar energy to recycle wastes and dead bodies back into structural materials and food—life maintains and increases its Earthly complexity over time. Like an aeons-long storm, it recycles the elements it needs into new life-forms and new generations, keeping, in a way that inanimate matter does not, the past *alive*—that is the word for it—in the present. Each of us is a kind of ancient relic, a living fossil harkening back to the early solar system. You may think you are twelve or thirty or sixty-two years old, but as a form of material entropy-producing "memory," you are approximately one-third the age of the universe.

With the origin of life a kind of rift opened up in the fabric of space-time. The great difference between the eons-long storm of life and a hurricane reducing the gradient between high and low pressure systems, of course, is that the solar gradient is so immense. Whereas the gravitational difference resolved by a whirlpool lasts seconds, life's creative destruction of the solar gradient has already lasted billions of years. Life's role, its purpose in the cosmic sense of thermodynamic change, is to reduce gradients. This is what we are doing when we turn oil in Earth's crust into a gas in the atmosphere. As heat in a heated cabin seems to obsessively search for ways to leave through the tiniest hole or crack, so life searches and finds ways to exploit energetic differences and funnel them into its own growth. But life, although it exhibits the same kind of manic purpose found in the inanimate behavior of "cool-seeking" heat, is trickier. With life, a strange thing happens on the way to gradient breakdown: the gradient-reducing systems, namely the open thermodynamic systems of life, self-destruct if they greedily deplete an available energy source. Thus life, though it obeys nature's meta-drive to bring everything to the perinirvana of uniform stasis, is forced to moderate itself in order not to destroy, along with the gradients, the very means of gradient breakdown.

We can herein trace a cosmic understanding of life, its desire, thought, and pleasure. Far from being aliens, islands of wisdom unconnected to the mute and unfeeling universe about us, we are energy transformers whose information-processing abilities ("wisdom," "thought," "planning") reflect nature's "craving"—her unconscious and, in us, conscious want—for reduction of ambient gradients. Our improbability and intricacy is an offshoot of the improbability of ambient gradients prior to us, such as the incredible unlikelihood of a universe with radiating suns. The "intelligence" of a tornado is linked to solving the problem of reducing barometric pressure differences; life, reducing the far greater solar gradient continuously for over three billion years now, has evolved far greater problem-solving abilities. On this view, the goal-setting, responsive tendencies of intelligent life do not look so much extraordinary as an ordinary part of an extraordinary universe. We are children of the sun, as Vernadsky said, entailing thermodynamic processes that are hardly unique. Ultimately, in our view, living growth and reproduction are manifestations of the more-than-living second law, which we believe will ultimately be seen as a major force in life's evolution—the most intriguing and open-ended example of change in the universe. And the pleasures of eating and sex, in us animals, are directly linked to our metabolic maintenance, growth, and the reproduction of that metabolic maintenance and energy-transforming growth in subsequent generations. Even those most human motivators—desires for food and shelter, sex and money, marriage and fame—seem to be far-reaching reflections of the cosmic tendencies of cream to disperse and heat to dissipate. Food maintains metabolism; shelter houses it; sex, money, and the rest increase the chances it will be present in our children, maintaining gradient-breaking forms of organization after our inevitable thermodynamic demise as individuals.

In the short run, during our individual lifetimes, we produce entropy by maintaining our identity, which necessarily entails the elimination of liquids, gases, and solids into the environment. In the long haul we assure entropy production by mating, thereby making new organisms like us that continue the special form of dissipation known as metabolism. From a cosmic vantage point, our interest in sex, eating, and preserving ourselves and others—life's greatest pleasures—is a by-product of the entropy production such activities engender; but life's self-repeating chemistry is but one of many gradient-breaking processes whose local coolness and complexity are made up for by the heat and entropy they beget, and whose seemingly transcendental designer-made intricacy is belied by the very Earthly mess they make in their midst.

A thermodynamic perspective of evolution teaches us that the only constant is change, and that this change is irreversible, inherent in the universe's original state of extreme improbability, and its mandate, as per the second law, is to unfold toward more probable distributions. At the same time we become aware of, if not almost hypnotized by, life's extreme chemical conservatism—a chemical conservatism permitted by a changing universe as spontaneously occurring open systems, able to reduce gradients, internally organizing structure and memory as they grow and export chaos to the area surrounding them. And of course this is what life, chemically and biologically, has done: it has grown, changing its surroundings, and then changed in the face of its changed surroundings in a running attempt to conserve its degradation abilities—to survive. As stated, with the origin of life, a kind of rift opened up in the fabric of space-time: maintaining metabolism, life repeats the process of its origins and its development, incarnating memory; the rest of the universe, meanwhile, continues on its merry forgetful way.

There are numerous implications. One is that we are literally living time capsules, museums in motion: by looking at modern cells, it should be possible to find vestiges not only of the metabolisms of the earliest life-forms but also of the ways by which life first evolved from complexifying gradient-breaking matter. Another implication, concerning the future, stands in dynamic contrast to the traditional twentieth-century "Copernican" or existentialist view of life as a mere speck in space and tick in time; it is that our tininess in the vastness of space is in fact in inverse proportion to our potential importance for the universe at large.

We know an oak grows from an acorn, but we do not know what happens when life originates in a universe. Life, despite its localization on the surface of a small planet orbiting a medium-sized star on the outskirts of a typical spiral galaxy (the Milky Way), is a most promising means of gradient reduction. With the evolution of humans and high technology there is raised the possibility that the entire universe, in the far future, may be grist for life's gradient-reducing mills. In which case life, despite its minuscule size, may be bound up with the entire future history of a changing universe. Whether you are religious or not, it is clear that life has really started something.

Chapter 22 Notes
1. Snow, 1963.
2. Carnot, reprinted 2005, also see Schneider and Sagan, 2006.
3. Schrödinger, 1946.
4. Vernadsky, 1998.
5. Prigogine, 1984.
6. Morowitz, 1979, 1993.

readings

Abram, D. 1996. *The Spell of the Sensuous*. New York: Pantheon Books.

Awramik, S. M. 1973. Stromatolites of the Gunflint Iron Formation. PhD dissertation., Harvard University Department of Geology, Cambridge MA.

Awramik, S. M., J. W. Schopf, and M. J. Walter. 1983. The Warrawoona microfossils. *Precambrian Research* 20:100.

Bada, J. L., and S. L. Miller. 1968. Ammonium ion concentration in the primitive ocean. *Science* 159:423–425.

Barghoorn, E. S. 1971. The oldest fossils. *Scientific American* 224:30–42.

Barghoorn, E. S., and J. W. Schopf. 1966. Microorganisms three billion years old from the Precambrian of South Africa. *Science* 152:758–763.

Barghoorn, E. S., and S. A. Tyler. 1965. Microorganisms from the Gunflint chert. *Science* 147:563–577.

Barks, C. and J. Moyne (Translators). 1999. *Open Secret: Versions of Rumi*. Boston: Shambhala; Athens GA: Maypop Books, 1987.

Bateson, W. B. 1928. "[S. Butler is the] most brilliant, and by far the most interesting of Darwin's opponents." From *Heredity and Variation in Modern Lights*. Cited in C. B. Bateson, ed. *William Bateson, F.R.S.* Cambridge UK: Cambridge University Press.

Bell, G. 1982. *The Masterpiece of Nature: The Evolution and Genetics of Sexuality*. Berkeley CA: University of California Press.

Berman, M. 1989. *Coming to Our Senses: Body and Spirit in the Hidden History of the West*. New York: Simon & Schuster, p. 239.

Botkin, D. B., and E. A. Keller. 1987. *Environmental Studies: The Earth as a Living Planet*. Columbus OH: Charles E. Merrill Pubs.

Botkin, D. B., P. A. Jordan, S. A. Dominski, H. D. Lowendorf, and G. E. Hutchinson. 1973. Sodium dynamics in a northern ecosystem. *Proceedings National Academy Science USA* 70:2745-2748.

Butler, S., 1914. *A First Year in Canterbury Settlement With Other Early Essays*. Edited by R. A. Streatfield. London: A. C. Fifield.

Canuto, V. M., J. S. Levine, T. R. Augustsson, and C. L. Imhoff. 1982. UV radiation from the young sun and oxygen and ozone levels in the prebiological palaeoatmosphere. *Nature* 296:816–820.

Carnot, S. 2005. *Reflections on the Motive Power of Fire: And other Papers on the Second Law of Thermodynamics*. New York: Dover Publications.

Case, E. 2008. Teaching taxonomy: how many kingdoms? *The American Biology Teacher* (National Association of Biology Teachers) in press.

Cloud, P. E., Jr. 1968. Atmospheric and hydrospheric evolution on the primitive Earth. *Science* 160:729-736.

———. 1983. The biosphere. *Scientific American* 249:176–189.

Darwin, C. 1871. (reprinted 1978) *The Descent of Man and Selection in Relation to Sex*. R. M. Hutchins, ed. Chicago: Britannica Great Books of the Western World.

———. 1859. *On the Origin of Species by Means of Natural Selection.* London: J. Murray.

Dawkins, R. 1976. *The Selfish Gene.* Oxford: Oxford University Press.

———. 1982. *The Extended Phenotype: The Gene as the Unit of Expression.* Oxford: W.H. Freeman

Doolittle, W. F. 1981. Is nature really motherly? *CoEvolution Quarterly* 29:58-63.

Duesberg, P. 1997. *Inventing the AIDS Virus.* Washington DC: Regnery Publishing.

Dyer, B. 1989. Symbiosis and organismal boundaries. *American Zoologist* 29:1085–1093.

Dyer, B. D. 2003 *A Field Guide to Bacteria.* Ithaca NY: Cornell University Press.

Dyer, B. D., and R. Obar, eds. 1985. *The Origin of Eukaryotic Cells: Benchmark Papers in Systematic and Evolutionary Biology.* New York: Van Nostrand Reinhold.

Dyson, G., 1999. *Darwin Among the Machines.* London: Penguin Books.

Edelman, G. M. 1985. Neural Darwinism: Population thinking and higher brain function. In M. Shafto, ed. *How We Know, Nobel Conference XX.* St. Peter, MN: Gustavus Adolphus College, pp. 1–30.

Egami, F. 1974. Minor elements and evolution. *Journal Molecular Evolution* 4:113–120.

Elias, N. 1978. *The Civilizing Process: The History of Manners.* Translated by E. Jephcott. New York: Urizen Books, pp. 252–253.

Ferris, J., and L. Nicodem. 1974. In K. Dose, S. W. Fox, C. A. Deborin, and T. E. Pavlovskaya, eds. *Origins of Life and Evolutionary Biochemistry.* New York: Plenum Press, pp. 107–117.

Fleck, L. 1979. *Genesis and Development of a Scientific Fact.* Chicago: University of Chicago Press.

Foucault, M. 1977. What is an author? In D. F. Bouchard (ed). *Language, Counter-Memory and Practice: Selected Essays and Interviews.* Ithaca NY: Cornell University Press, p. 124.

Futuyma, D. J. 1985. Is Darwinism dead? *The Science Teacher* 52:16–21.

Giusto, J. P., and L. Margulis. 1981. Karyotypic fission theory and the evolution of old world monkeys and apes. *Biosystems* 13:267–302.

Gold, P. E. 1987. Sweet memories. *American Scientist* 75:151–155.

Goody, R., and J. C. G. Walker. 1972. *Atmospheres.* Englewood Cliffs NJ: Prentice Hall.

Greenberg, N, and P. D. MacLean, eds., 1978. *Behavior and Neurology of Lizards, an Interdisciplinary Colloquium* (Rockville, MD: U.S. Department of Health, Education, and Welfare, Public Health Service, Alcohol, Drug Abuse, and Mental Health Administration, National Institute of Mental Health), DHEW Publication no. (ADM) 77–491.

Gregory, P. H. 1973. *Microbiology of the Atmosphere,* 2nd edition. New York: John Wiley.

Grinevald, J. 1988. A history of the idea of the biosphere. In P. Bunyard and E. Goldsmith, eds. *Gaia: The Thesis, the Mechanisms and the Implications.*

Proceedings of the First Annual Camelford Conference on the Implications of the Gaia Hypothesis. Cornwall, UK: Quintrell & Co. Reprinted 1995 in Bunyard, P., ed. *Gaia in Action: Science of the Living Earth.* Edinburgh UK: Floris Books.

Habermas, J. 1987. *The Philosophical Discourse of Modernity.* Translated by F. Lawrence. Cambridge MA: MIT Press

Hall, J. L., and L. Margulis. 2008 (forthcoming). Genetic contribution of spirochetes to the eukaryotic lineage: A novel bioinformatic analysis. *Proceedings of the National Academy of Sciences.*

Hall, J. L., Z. Ramanis, and D. J. L. Luck. 1989. Basal body/centriolar DNA: Molecular genetic studies in *Chlamydomonas. Cell* 59:121–132.

Hayden, D. 2003. *Pox: Genius, Madness, and the Mysteries of Syphilis.* New York: Basic Books.

Heim, M. 1993. *The Metaphysics of Virtual Reality.* New York: Oxford University Press.

Hughes, J. D. 1983. Gaia: An ancient view of our planet. *The Ecologist: Journal Post-Industrial Age* 13:54–60.

Hutchinson, G. E. 1954. *Biogeochemistry of Vertebrate Excretion.* New York: American Museum of Natural History.

Huxley, J. 1912. *The Individual in the Animal Kingdom.* New York: G.P. Putnam's Sons (reprinted 1994 New Haven CT: Ox Bow Press).

Jacob, F. 1973. *The Logic of Life: A History of Heredity.* New York: Pantheon, p. 25.

Jonas, Hans, and Elenore Jonas. 1966. *The Phenomenon of Life: Toward a Philosophical Biology.* Evanston IL: Northwestern University Press.

Kaveski, S., D. C. Mehos, and L. Margulis. 1983. There's no such thing as a one-celled plant or animal. *The Science Teacher* 50:34–36, 41–43.

Kellogg, W. W., and S. H. Schneider. 1974. Climate stabilization: For better or for worse? *Science* 186:1163–1172.

Kendrick, B. 2001. *The Fifth Kingdom,* 3rd edition. Victoria BC: Mycologue Publications.

Khakhina, L. N. 1992. *Concepts of Symbiogenesis: A Historical and Critical Study of the Research of Russian Botanists.* M. A. McMenamin and L. Margulis, eds. New Haven CT: Yale University Press.

Klossowski, P. 1991. *Sade My Neighbor.* Translated by A. Lingis. Evanston IL: Northwestern University Press.

Knoll, Andrew. 2003. *Life on a Young Planet.* Princeton NJ: Princeton University Press.

Kolnicki, R. L. 1999. Karyotypic fission theory applied, kinetochore reproduction and lemur evolution. *Symbiosis* 26:123–141.

———. 2000. Kinetochore reproduction in animal evolution: Cell biological explanation of karyotypic fission theory. *Proceedings National Academy Science* 97:9493–9497.

Kornstein, S., and A. Clayton. 2002. *Women's Mental Health: A Comprehensive Textbook.* New York: The Guilford Press.

Koshland, D. E., Jr. 1992. A response regulator model in a simple sensory system. *Science* 196:1055–1056.

Kozo-Polyansky, B. M. 1924. *Symbiogenesis: A new principle of evolution* (in Russian). Translated with commentary by Victor Fet. White River Junction, VT: Sciencewriters–Chelsea Green Publishing Company. (2008, forthcoming).

Lacan, J. 1977. The mirror stage as formative in the function of the I. In *Écrits: A Selection*. Translated by A. Sheridan. New York: W. W. Norton, pp. 1–7.

Lapo, A. V. 1988. *Traces of Bygone Biospheres*. Moscow: MIR. Translated by V. Purto. Oracle, AZ: Synergistic Press.

Leenhardt, M. 1979. *Do Kamo*. Translated by B.M. Gluati. Chicago: University of Chicago, p. 22.

Legato, M. J., 1998. *Women's Health Not for Women Only, International Journal of Fertility and Women's Medicine*, 43: 65–72.

Lovelock, J. E. 1972. Gaia as seen through the atmosphere. *Atmosphere Environment* 6:579–580.

———. 1979. *Gaia: A New Look at Life on Earth*. Oxford UK: Oxford University Press.

———. 1983a. Gaia as seen through the atmosphere. In P. Westbroek and E. W: de Joeng, eds. *The Fourth International Symposium on Biomineralization*. Dordrecht, Holland: Reidel.

———. 1983b. Daisy World: A cybernetic proof of the Gaia hypothesis. *CoEvolution Quarterly* 31:66–72.

———. 1988. *The Ages of Gaia*. New York: W. W. Norton.

———. 2006. *Revenge of Gaia*. New York: Basic Books.

———. 2006. Making peace with Gaia. *Resurgence* 238:59-61.

Lovelock, J. E., and J. P. Lodge. 1972. Oxygen in the contemporary atmosphere. *Atmosphere Environment* 6:575–578.

Lovelock, J. E., and L. Margulis. 1974a. Atmospheric homeostasis by and for the biosphere: The Gaia hypothesis. *Tellus* 26:2–10.

Lovelock, J. E., and L. Margulis. 1974b. Homeostatic tendencies of the Earth's atmosphere. *Origins of Life* 1:12–22.

Lovelock, J. E., and L. Margulis. 1976. Is Mars a spaceship too? *Natural History Magazine* 85:86–90.

Madigan, M., J. Martinko, and J. Parker. 2003. *Brock's Biology of Microorganisms*. Englewood Cliffs NJ: Prentice Hall.

Margulis, L. 1990 Words as battlecries: Symbiogenesis and the new field of endocytobiology. *BioScience* 40:673–677.

———. 1993. *Symbiosis in Cell Evolution*, 2nd edition. New York: W.H. Freeman.

———. 1998 *Symbiotic Planet: A New Look at Evolution*. New York: Basic Books.

———. 2005. Jointed threads. *Natural History Magazine*. June, pp. 28–32.

———. 2007. *Luminous Fish: Tales of Science and Love*. White River Junction VT: Chelsea Green Publishing Company.

Margulis, L., and M. J. Chapman. 2008 (forthcoming). *Kingdoms and Domains: An Illustrated Guide to the Phyla of Life on Earth*, 4th edition. New York: Academic Press.

Margulis, L. and Dolan, M. F. *2002 Early Life: Evolution on the Precambrian Earth*. 2nd ed. Sudbury MA: Jones and Bartlett.

Margulis, L., and J. E. Lovelock. 1974. Biological modulation of the Earth's atmosphere. *Icarus* 21:471–489.

Margulis, L. and Punset, E. eds. 2007. *Mind, Life, and Universe: Conversations with Great Scientists of Our Time*. White River Junction VT: Chelsea Green Publishing.

Margulis, L., and D. Sagan. 1991. *Origins of Sex*. New Haven CT: Yale University Press.

Margulis, L., and D. Sagan. 1997. *Microcosmos: Four Billion Years of Evolution from Our Bacterial Ancestors*. Berkeley CA: University of California Press.

Margulis, L., and D. Sagan. 2000 *What Is Life?* Berkeley CA: University of California Press.

Margulis, L., and D. Sagan. *2002 Acquiring Genomes: A Theory of the Origins of Species*. New York: Basic Books.

Margulis, L. and J. MacAllister. 2005. *Eukaryosis: Origin of Eukaryotic Cells*. Video/DVD. Amherst, MA.

Margulis, L., J. O. Corliss, M. Melkonian, and D. J. Chapman, eds. 1990. *Handbook of Protoctista: Structure, Cultivation, Habitats and Life Cycles of the Eukaryotic Microorganisms and their Descendants Exclusive of Animals, Plants and Fungi*. Sudbury MA: Jones and Bartlett. (Second edition forthcoming 2009.)

Margulis, L., H. I. McKhahn, and L. Olendzenski. 1993. *Illustrated glossary of the Protoctista*. Sudbury MA: Jones and Bartlett.

Margulis, L., J. C. G. Walker, and M. B. Rambler. 1976. A reassessment of the roles of oxygen and ultraviolet light in Precambrian evolution. *Nature* 264:620–624.

Margulis, L., J. Z. Jorgensen, M. F. Dolan, R. Kolchinsky, F. A. Rainey and S-C. Lo. 1998. The *Arthromitus* stage of *Bacillus cereus*: Intestinal symbionts of animals. *Proceedings National Academy Science* 95:1236–1241.

Margulis, L., M. Chapman, R. Guerrero, and J. Hall. 2006 The Last Eukaryotic Common Ancestor (LECA): Acquisition of cytoskeletal motility from aerotolerant spirochetes in the Proterozoic eon. *Proceedings National Academy Science*. 103:13080–13085.

Maturana, H. R., and F. J. Varela, 1973. Autopoiesis: The organization of the living. In H. R. Maturana and F. J. Varela, eds. *Autopoiesis and Cognition*. Boston MA: D. Reidel.

Miyashita, Y., and H.S. Chang. 1988. Neuronal correlate of pictorial short-term memory in the primate temporal cortex. *Nature* 331:68-70.

McCarthy, B., and E. McCarthy, 1998. *Male Sexual Awareness: Increasing Sexual Satisfaction*, New York: Carroll and Graf.

Morowitz, Harold J., 1979. *Energy Flow in Biology*, New Haven CT: Ox Bow Press
———. 1993. *Beginnings of Cellular Life: Metabolism Recapitulates Biogenesis*. New Haven CT: Yale University Press.

Morley, H. 1880. *Memoirs of St. Bartholomew's Fair*. London: Chatto and Windus. 231.

Newman, M. J. 1980. Evolution of the solar "constant." In C. Ponnamperuma and L. Margulis, eds. *Limits to Life*. Dordrecht, Holland: Reidel.

Oster, L. 1973. *Modern Astronomy*. San Francisco: Holden Day.

Pagel, W. 1951. William Harvey and the purpose of circulation. *Isis* 42:22–38.

Parker, G. A. 1970. Sperm competition and its evolutionary consequences in the insects. *Biological Reviews* 45: 525–567.

Persinger, M. 1987. *Neuropsychological Bases of God Beliefs*. Westport CT: Praeger Publishers.

Pert, C., and N. Griffiths-Marriott. 1988. Bodymind. *Woman of Power* 11:22–25.

Poundstone, W. 1999. *Carl Sagan: A Life in the Cosmos*. New York: Henry Holt.

Prigogine, I. 1981 *From Being to Becoming* New York: W. H. Freeman

Prigogine, I., and Stengers, I. 1984. *Order out of Chaos: Man's new Dialogue with Nature*. New York: Bantam Books.

Rasool, I., ed. 1974. *The Lower Atmosphere*. New York: Plenum Press.

Roszak, T., 1999. *The Gendered Atom: Reflections on the Sexual Psychology of Science*. Nanaimo: Conari Press.

Ryan, F. P. 2007. Viruses as symbionts. *Symbiosis* 44:1–100.

Sagan, C. 1970. *Planets and Explorations*. Condon Lectures. Eugene OR: Oregon State System.

Sagan, C, and G. Mullen. 1972. Earth and Mars: Evolution of atmosphere and surface temperatures. *Science* 177:52–56.

Sagan, D. 1987. Biosphere II: Meeting ground for ecology and technology. *Environmentalist* 7:271–281.

———. 1990a. *Biospheres. Metamorphosis of Planet Earth*. New York: McGraw-Hill.

———. 1990b. What Narcissus saw: The Oceanic "I"/"eye." In *Speculations: The Reality Club* 1. J. Brockman, ed. Englewood Cliffs NJ: Prentice Hall, pp. 245–266.

———. 2007. *Notes from the Holocene: A brief history of the future*. White River Junction VT: Chelsea Green Publishing Company.

Sagan, D., and L. Margulis. 1993. *Garden of Microbial Delights: A Practical Guide to the Subvisible World*. Dubuque IA: Kendall-Hunt Publishers.

Schneider, S. H. , J. R. Miller, E. Crist, and P. J. Boston, eds. 2004. *Scientists Debate Gaia: The Next Century*. Cambridge MA: MIT Press.

Schneider, E., and D. Sagan. 2006. *Into the Cool: Energy Flow, Thermodynamics, and Life*. Chicago: University of Chicago Press.

Schopf, J. W. 1970. Precambrian microorganisms and evolutionary events prior to the origin of vascular plants. *Biological Reviews* 45:31–352.

Schopf, J. W. (ed). 1983. *Precambrian Paleobiology Research Group Report*. Princeton NJ: Princeton University Press.

Schrödinger, E. 1946. *What is Life?* New York: Macmillan.

Shukla, J., and Y. Mintz. 1982. Influence of the land-surface evapotranspiration on the Earth's climate. *Science* 215:1498–1501.

Simmons, J. 1996. *The Scientific 100: A Ranking of Scientists Past and Present.* Seacaucus NJ: Carol Publishing.

Simpson, G. G. 1960. In S. Tax, ed. *Evolution after Darwin.* Vol. 1. Chicago: University of Chicago Press, pp. 177–180.

Smil, V. 2003. *The Earth's Biosphere.* Cambridge MA: MIT Press.

Smith, R. L. 1984. Human Sperm Competition. In *Sperm Competition and the Evolution of Animal Mating Systems.* Robert Smith, ed. Orlando FL: Academic Press, 601–653.

Snell, B. 1960. *The Discovery of the Mind.* Translated by T. C Rosenmeyer. New York: Harper Torchbooks.

Snow, C. P., 1964. *The Two Cultures: A Second Look.* Cambridge UK: Cambridge University Press.

Sonea, S., and L. G. Mathieu, 2002. *Prokaryotology: A Coherent View.* Montréal: Les Presses de L'Université de Montréal.

Stadtman, T. 1974. Selenium biochemistry. *Science* 183:915–922.

Thomas, Lewis. 1978. *Lives of a Cell: Notes of a Biology Watcher.* New York: Penguin.

Todd, N. 1970. Karyotypic fissioning and carnivore evolution. *Journal Theoretical Biology* 26:445–480.

Uexküll, J. von 1926. *Theoretical Biology.* Translated by D. L. MacKinnon. *International Library of Psychology, Philosophy and Scientific Method.* London: Kegan Paul, Trench, Trubner & Co.

Vernadsky, V. I. 1929. *La Biosphère.* Abbreviated version translated as *The Biosphere.* 1986. Oracle, AZ: Synergetic Press. Translation from the Russian 1926 edition with commentary by Mark McMenamin, Vernadsky, V. I. 1998, *The Biosphere* Copernicus Books, New York: Springer-Verlag.

Waddington, C. H. 1976. Concluding remarks. In E. Jantsch and C.H. Waddington, eds. *Evolution and Consciousness.* Reading, MA: Addison Wesley, pp. 243–250.

Wakeford, T., and M. Walter, eds. 1995. *Science for the Earth.* Chichester, UK: John Wiley & Sons, pp. 19–37.

Wallin, I. E. 1927. *Symbionticism and the Origin of Species.* Baltimore: Williams & Wilkins.

Walter, M. R., ed., 1976. *Stromatolites: Developments in Sedimentology.* Vol. 20. New York: Elsevier.

Watson, A. J., and J. E. Lovelock. 1983. Biological homeostasis of the global environment: The parable of "Daisy World." *Tellus* 35:284–289.

Watson, A. J., J. E. Lovelock, and L. Margulis. 1978. Methanogenesis, fires, and the regulation of atmospheric oxygen. *BioSystems* 10:293–298.

Westbroek, P. 1991. Life as a Geological Force. New York: W. W. Norton. (Second edition forthcoming 2008.)

Wilson, E. O. 1992. *The Insect Societies,* 2nd edition. Cambridge MA: Harvard University Press.

figure and table list

figures

Chapter Openers by Abrah Griggs
Mnemosyne   1
Chimera   27
Eros   109
Gaea   153

1.1   Sagan wedding   4
3.1   *Navicula*   21
3.2   Branched *Arthromitus* (micrograph)   23
3.3   Branched *Arthromitus* (drawing)   23
3.4   *Stigmatella*   24
4.1   *Stephanonympha*   34
4.2   *Rhodophyta*   34
5.1   Sperm tails   40
5.2   Spirochetes become undulipodia (drawing)   41
6.1   *Mixotricha paradoxa*   42
6.2   *Canaleparolina*   46
7.1   Spirochetes become undulipodia (diagram)   50
7.2   *Karyomastigont* formation   51
8.1   Friedrich Nietzsche   58
8.2   Pox inspection of the ladies of the night   59
8.3   *Spirosymplokos deltaeiberi*, life history   66
8.4   *Spirosymplokos deltaeiberi*, round bodies   67
8.5   *Cristispira* (drawing)   68
9.1   Kefir microbes/"Muhammed pellets"   74
10.1 Intellectual ferment at UMass   77
10.2 Printing press with monkeys   82
10.3 Samuel Butler on computer screen   83
10.4 Termite: *Pterotermes occidentis*   85
13.1 Sex in ciliates   113
13.2 Bacterial sex without reproduction   116
13.3 *Paramecia* mating   117
13.4 Abortive cannibalism   117
13.5 Mating *trichonympha*   119
14.1 Whale penis   124
15.1 Fetal genitalia   141
16.1 Zany, clown of his Mountebank   148
17.1 Oceanus macromicrocosmicus   158

17.2   Earth's atmosphere with life   165
17.3   Earth's atmosphere now if there were no life   166
17.4   Earth in the Archean eon   168
17.5   Earth in the Proterozoic eon   168
17.6   Metabolism in an anoxic world   169
17.7   Metabolism in an oxic world   169
19.1   *Emiliania* from space   187
19.2   *Emiliania huxleyi* with coccoliths   189
19.3   Michrobial mat cutaway   193
20.1   Amoeba cannibalism   202

tables

9.1   Kefir: components live microbes   73
17.1   Reactive gases in the atmosphere   161
17.2   Atmospheric gases in disequilibrium   162
17.3   Naturally limiting biological elements   163

index

*Entries in **bold italics** refer to figures and tables.

abortive cannibalism, 110, 177
Abram, David, 88
acidified lakes, 190
acid rain precursors, 190
advancement, 86
aerotolerant aerobes, 167
agglutinating *foraminifera*, 81
AIDS patients, 62
AIDS virus, 24–25, 63
Alexander, Leone, 2
Alexander, Morris, 2
algae, 43
Allen, Joseph, 218
Amazon rainforests, 229
ammonia atmospheres, 164
Amoeba cannibalism, 202
Amoeba proteus, 201
amygdala, 105
anaerobic bacteria, 191
anaerobic cycles, 167
Andreae, Meinrat, 188–189
animal evolution focus, 215
animal kingdom, 214
animals, treatment of, 96
animistic worldview, 221
Anolis lizard, 133
anoxic world, 169
anthropocentrism, 183
Aplysia (sea hare), 48
Arcana Naturae Detecta, 52
archaebacteria (archaea), 35, 51, 161
Archean eon Earth, 168
Aristotle, 223
Arouet, François-Marie (Voltaire), 198
arsphenamine, 64
Arthromitus (multicellular bacteria), 23
asexual reproduction, 112
Ask the Dust (Fante), 149
aspirin, 143
atmospheric chemistry of Earth, 162, 178
atmospheric gases, 161-162
atmospheric homeostasis, 163–164, 170
atmospheric methane, 174

atmospheric studies, 159–160
atomistic model of organismic identity, 21
Australian flightless birds, 38
Australopithecus africanus, 127
autopoiesis, 16, 228
autopoietic entities, 22, 23
awareness, 48, 53

bacteria. *See also* cyanobacteria
   anaerobic, 191
   as ancestors to humans, 37
   archaebacteria, 51, 161
   benefits provided by, 37
   chemotactic, 202
   communities of, 32
   cyst-forming myxobacteria, 22
   disease, 37
   E. coli, 202
   endosymbiotic, 17
   fermenting, 191
   gene exchanges by, 31
   as germs, 36
   gram negative, 60
   in humans, 37, 45
   intestinal, 45
   mergers with and by, 31
   multicellular, 22
   photosynthetic, 179
   respiring, 92
   sex, 113
bacteria kingdom, 33–34
bacterial cell, 22
bacterial consciousness, 37
bacterial mergers, 38, 40
bacterial sex, 113–115, 120
Bartholinus, Thomas, 157
Bateson, Gregory, 203
Bateson, Mary Catherine, 140–141
Bateson, William, 84
"BBC Theorem", 180
Beauchemin, Ginette, 72
Beggiatoa, 187
behavioral neuroscience, 104
Belichenko, Andrei, 65

biological evolution, 224
Biosfera (The Biosphere) (Vernadsky), 81
biosphere
    atmosphere as extension of, 159
    feedback system, 162
    futures of, 90–92
    as home vs. body, 34 ˙
    human dependence on, 89
    survivablility of, 91–92
    temperature homeorrhesis of, 177
Biosphere II, 216–217
biospheres
    communication between, 218
    development of, 219–220
biospheric individuality, 18
biparental sex, 120
birds, 134
Blake, William, 96
blood-brain barrier, 24, 65
blue-green algae. *See* prokaryotes
the body, 17, 19
body size, 127
Bohr, Niels, 218
Boltzmann Ludwig, 224
boojum tree *(Fouquieria columnaris* or
    Idria columnaris), 81
books, 97
Borrelia burgdorferi (Lyme disease
    spirochete), 60, 62, 65, 69
Bosch, Hieronymous, 20
brains
    electromagnetic (EM) stimulation of,
        106, 110
    electromagnetic algorithms under-
        lying specific experiences in, 104
    and ESP, 103
    gender differentiation, 140–142
    as mind, 48–49
    as representing outside world, 103
Brand, Stewart, 157
breasts, 123, 130, 142
Brorson, Sverre-Henning, 68–69
Brorson Oystein, 68–69
Bukowski, Charles, 149
Burckhard, Jacob, 57
Butler, Samuel, 25, *83*, 84, 203–205
Byrne, David, 223
Calvin, William, 205
Canale-Parola, Ercole, 46
Canaleparolina, 46, 53

Candida albicans, 147–148, 150–152
cannibalism, 31, 117
Cantor, Charles, 49
carbon-dioxide-removing communities,
    179
cardiolipin, 64
Carnot, Nicolas Leonard Sadi, 227
Carson, Johny, 5, 8
Cartesian anthropocentrism, 206
Cartesian dualism, 195
Cartesian license, 196–197, 198
Cartesian tradition, 203
cell movement, 52
cervical cancer, 147
change
    inevitability of, 224, 231
    as intrinsic to life, 233
Charles, King of France, 61
chemotactic bacteria, 202
chimera, 27–28
chimpanzees
    in estrus, 123, 128
    genitalia comparisons, 126
    and humans, 128
Chlamydomonas, 50
chloroplasts, 43, 44
Chromatium, 187
chromosome counts, 49
ciliate sex, 113
Clarke, Arthur C., 93
classification systems, 29, 33
Cleveland, L. R., 45, 137
climax communities, 80
cloning, 149
clowns, 147, *148*, 150–152
coccolithophores, 213
coevolution
    with machines, 77
    with microbes, 95
    with species creation, 95
    species creation through, 95
CoEvolution Quarterly, 157
coexistence, 96
Cohen, Yehuda, 192
colonization of Mars, 88
Columbus, Christopher, 61
communication, 78–79
complexity, 225, 228, 229, 232
composite symbiotic complex, 24
conscious entities, 38

consciousness, 38, 48
consumption, 52
controlled ecosystems, 218
convergence, 91
Copenhagen interpretation of atomic structure, 20
Copromonas, 201
correspondence theory of reality, 19
Cosmos, 9, 13
cows, 95
creative destruction of solar gradient, 234–235
Cretaceous catastrophe, 92
Cristispira, 68
cultural evolution, 93
culture and language, 144
cyanobacteria, 43, 161, 191–192, 231, 232, 234. *See also* prokaryotes
Cyanobacterial communities, 80–81
cybernetic systems, 175–176
cybersymbiosis, 100
cyst-forming myxobacteria, 22, 24
cysts, 68–69

Daisyworld model, 172, 175–177
Darwin, Charles, 83, 123, 125, 128–129, 157
Darwinian evolution
  modern synthesis of, 214
  unit of selection in, 21
Das Energi (Williams, Paul), 101
da Vinci, Leonardo, 186
Davis, Bette, 123
Dawkins, Richard, 180
death
  described, 70
  and individuality, 74
  as sexually transmitted disease, 146
death genes, 71
Demon-Haunted World (Sagan, Carl and Ann Druyan), 102
depression, 141, 144
Derrida, Jacques, 209
Descartes, René, 195, 206
Dialogue Concerning the Two Chief World Systems (Galileo), 157, 197
Diaz de Isla, Ruiz, 61
dichotomies, 29
Dick, Philip K., 101, 103
Dickinson, Emily, v, xi

dimethyl sulfide, 188, 213
dinosaurs, 92, 133
disease bacteria, 37
disintegrative relations, 24
dissipative structures, 233
dizygotic (fraternal) twins, 125
DNA
  in mitochondria, 43
  in nucleus, 43
  purification of, 49–50
DNA repair, 114
Dolan, Michael, 37
Doolittle, W. Ford, 180
dormancy, 67–68
Dostoyevsky, Fyodor, 105
dreams, 102
Drum, Ryan, 99
Druyan, Ann, 9, 12
Duesberg, Peter, 57, 63
Dyer, Betsey, 53

Earth. *See also* biosphere
  atmosphere as circulatory system, 157–159
  atmospheric chemistry. of, 178
  atmosphere with life, 165
  atmosphere without life, 166
  biosphere of, 155–156
  as body, 18
  element cycles, 185
  as global metaorganism (alive), 209–210
  intelligent life on, 102
  magnetic field of, 106
  oxidation of, 192
  oxygen on, 165–166
  redox potential of, 232
  regulation by and for the biota, 172
  reproduction of, 216–218
  and solar luminosity, 173
  at Sun's life end, 94
  as superorganism, 213
  temperature modulation on, 213–214
  temperature regulation, 173
  temperatures on, 165, 231
  thermostasis of, 181
Earth lights, 106
earthquakes and UFO sightings, 106–107
Ecce Homo (Nietzsche), 61

E. coli bacteria, 202
Economics (Xenophon), 182
Eddington, Arthur, 227
Edelman, Gerald, 54–55, 205
educational system, 6
egointruder presence, 105
Ehrenberg, Christian Gotfried, 200
Eiser, Otto, 61
electromagnetic (EM) stimulation of
   brain, 106, 110
element cycles, 185–186
Elias, Norbert, 17
Eliot, T. S., 155
Emiliania from space, 187
Emiliania coccoliths, 189
entropy, 227
entropy production, 236
entropy reduction, 154
environmentalists, 183
epileptic seizures, 105
Erewhon (Butler), 84
error correction, 176
Escherichia coli, 114
ESP, 103
Esteve, Isabel, 68
estrogen, 143
estrus, 128, 130
eternity, 223
Euglena gracilis, 29
eukaryotes, 34
eukaryotic organisms, 22
evapotranspiration, 174
evapotranspiration of trees, 231
everywhereness of life, 200
evolution. *See also* coevolution
   accelerated nature of, 93
   animal evolution focus, 215
   biological, 224
   cultural, 93
   from forced fertilization, 97
   from *Homo sapiens*, 96
   ladder of being notion, 228
   linear view of, 30
   of machines, 85
   via plasmids, 97
evolutionary history
   of nucleated cells, 66
   tracing, 39
evolutionary innovation, 91
evolutionary transition, 38

evolutionary vestiges, 38–39
Explorer 10, 114
extraterrestial endeavors, 183
extraterrestial intelligence, 102

facultative aerobes, 167
faith, 208–209
family vs. career, 5–6
Famintsyn, Andrey S., 44, 45
feedback loops, 164, 176
female breasts, 123, 130, 142
Ferdinand, King of Spain, 61
fermenting bacteria, 191
fertilization factors, 125
fertilization-meiosis cycle, 75
fetal genitalia, 141
Field Notes (radio program), 36
Fleck, Ludwik 16, 62
Flo (chimpanzee), 128
Folsome, Clair, 218
foraminifera (forams), 81, 201
forced fertilization, 97
fossil records, 165
Fracastoro, Girolamo, 62
Franklin, Benjamin, 76
Freud, Sigmund, 131
Frucht, Bill, 102
Fuller, Buckminster, 92
Fuller, Robert, 190
fungi kingdom, 34–35

Gaea (Gaia), 153–154, 157
Gaia hypothesis
   biological worldview of, 172
   defined, 159
   described, 177
   effects of, 18
   impact of, 159
   postulates a live Earth, 213
Gaian control mechanisms, 183
Galapagos (Vonnegut), 149
Galileo, 157, 196–197
gametes, 144–145
Garrels, Robert M., 194
Gast, Peter, 57
gender development, 32
gender differentiation, 140–145
gene exchanges by bacteria, 31
genesis of the first cell, 180
genetic doubling, 111

genetic recombination, 114–115
genitalia comparisons, 125–126, 141
genome(s)
  of *Borrelia burgdorferi*, 65
  of *Leptospira*, 65
  numbers of, 43
  in protists, 43
  of *Treponema pallidum*, 65
  triple, in plants, 43
geomagnetic phenomena, 104
geophysiology, 194
global self-regulation, 172
global skeleton, 188
God, 14, 195, 197, 198, 203, 204, 209
Golding, William, 154, 155, 173
Gomel, Abe, 72
Goodall, Jane, 128
Goody, R., 159
Gore, Al, 228
Gorilla
  genitalia comparisons, 123
  sexual behavior, 128
  sperm-competition avoidance,
    126–127
gram negative bacteria, 60
Griffiths-Marriott, Nancy, 24
Grimstone, A.V., 45
Grinspoon, Lester, 144
group selection, 215, 216
Guerrero, Ricardo, 37, 68, 69

Habermas, Jürgen, 16
habitat
  boojum tree (*Fouquieria columnaris*
    or *Idria columnaris*) as, 81
  Lycium as, 81
  redwood tree as, 81
  saguaro cactus as, 81
hairy mastigotes, 137
Hall, John, 37, 49
Hansen, Joseph, 218
haploid cells, 115
harems, 126–127
Harvey, William, 157
Hawking, Stephen, 149
Hayden, Deborah, 57–58, 61, 69
Heim, Michael, 82
Herakleitos, 224
hierarchy theory, 21
Hoffmann, Erich, 62

Hogg, John, 43
Holser, William, 187
homeorrhetic systems, 176
Homo erectus, 78, 127
Homo ergaster, 78
Homo habilis, 127
Homo photosyntheticus, 99
Homo sapiens
  evolution from, 96
  extinction of, 93
  future of, 98–100
hormones, 142
Hughes, J. D., 182
human brain, 131
human immunological retrovirus
  (HIV), 57
human intellect, 181
human intelligent consciousness, 32
humanity
  as apotheosis of life, 90
  forecasts for, 90
human-machine communities, 87
human papilloma virus (HPV), 146,
  147
human reproduction, 121
humans
  bacteria in, 37
  and chimpanzees, 128
  composition of, 28
  extinction of sex in, 146
  genitalia comparisons, 126–128, 141
  intestinal bacteria in, 45
  isolated evolution of, 216
  mitochondria in, 45
  as obligate technobes, 90
  as pioneer species, 79
  reliance on machines, 87
hunter-gatherers, 129
Hutchinson, G. E., 174
Huxley, Julian, 18, 20
hydra, 27–28
hydrogen sulfide, 38, 233

the I, 18, 19
identical twins, 54–55
immune system (body) vs. nervous
  system (mind), 24
incest taboos, 129
individuality and death, 74
inevitability of change, 231

inner-ear cells, 39
integrative mergers, 24
intelligent design, 228
intersexual selection, 123, 125
intestinal bacteria, 45
intrasexual selection, 123, 125
intuition, 140
Inventing the AIDS Virus (Duesberg), 63, 239
in vitro fertilization, 149
isolation growth, 65

Jackson, Wes, 217
James, Henry, Jr., 105
James, Henry, Sr., 105
James, William, 105
Jeffers, Robinson, 102
Joan of Arc, 105
Johnson, Russell, 62
Jung, Carl, 212

Kant, Immanuel, 17
Kantian challenge, 225
Karyomastigont, 51
karyotypic fissioning, 99
Kaufmann, Stuart, 228
kefir, 72–74
Kierkegaard, Søren, 208
kinetochores, 115–116, 118
kinetosomes, 50
kingdoms, 33
Kirchner, James W., 220
knowledge, 48
Koren, Stanley, 105
Koren helmet, 105
Kornstein, Susan G., 141
Koshland, Daniel, 202
Kozo-Polyansky, Boris, 44
Kundera, Milan, 209

Lacan, Jacques, 17–18, 145
Lactobacillus ("acidophilus"), 150–151
language, 140
Late Night Thoughts On Listening to Mahler's Ninth Symphony (Thomas), 36–37
learning
  survival by, 78
  of technology, 78

Legato, Marianne, 143
Leidy, Joseph, 44
Leoniceno, Nicolò, 62
Leptospira spirochetes, 60, 65
leptospirosis, 60
Le Roy, Edouard, 200
lichen, 24
life
  Cartesian tradition on, 203
  everywhereness of, 200
  as fossil memory, 234
  as gradient reduction, 237
  incorporation of environment by, 77, 80
  interdependence of, 35
  as open system, 231
  understanding of, 235–236
light-sensitive cells, 39
limiting elements, 163
linear evolution view, 30
living matter as geological force, 199
long-term memory, 54
Lovelock, James E., 18, 154, 172–178, 181, 183, 189, 199, 201
Luck, David, 49
Lucy (fossil), 129
luminous balls, 106–107
Lycium, 81
Lyme disease, 60, 65, 69, 62

machine-humans, 88
machines
  coevolution with, 77
  evolution of, 85
  expansion of, 84–85
  human reliance on, 87
  as organelles of technological society, 85
MacLean, Paul D., 131–132, 135
magnetism theory, 101
malarial fever, 52
male and female designations, 118
males, competition among, 124
Margulis, Lynn (LM)
  on Carl Sagan's jealousy, 13
  family vs. career, 5–6
  Gaia hypothesis concept, 173
  lifelong goals of, 3
  marriage of, 2

marijuana, 144
marriage, 128
Mars
  atmospheric conditions on, 178
  carbon dioxide atmosphere, 231
  colonization of, 88
  dust storms on, 9
  life detection on, 154
  mass extinctions, 92
Masson, Jeffrey Moussaieff, 13
Mastotermes darwiniensis, 45
Mastigias, 99
Mastotermes darwiniensis, 43, 44
Mating Trichonympha, 119
Mathieu, Leo, 138, 146
Mayr, Christa, 101, 102
Mayr, Ernst, 101, 102
McCarthy, B., 127–128
McCarthy, E., 127–128
meiosis, 72, 120
meiotic chromosome reduction, 115–116
meiotic sex, 72, 115–118, 136–137
  origins of, 137
  and programmed death, 71
  and tissue-level multicellularity, 119
Melanesians, 17
membrane-bounded cell, 21–22
membranes, 17
memory, 48
Merezhkovsky, Konstantin Sergeeivich, 43, 44, 45
metabolic transformation, 33
metaphysical realism, 221
methanogens, 179
Mexican eyeless cavefish, 38
microaerophils, 167
microbes
  and carbon dioxide gas levels, 179
  choices made by, 201–202
  coevolution with, 95
  essential nature of, 32
  as germs, 29
  reproduction of, 146
microbial communities, integration of, 28
microbial growth and heat-modulating gases, 181
microbial mats, 67, 136, 192
microbial mind, 39

microbiogeochemistry, 194
Micrococcus radiodurans (Deinococcus), 93
Microcoleus chthonoplastes, 193
Microcosmos, 13
microscope, 29
minaturization, 97
mind and body, 205
mindbody-bodymind, 25
mind-brain processes, 53
mind phenomena, 50–51
miniaturized ecosystems, 218
mitochondria
  DNA in, 43
  genes for photosynthesis in, 138
  in humans, 31, 45
  immortality of, 28
mitosis, 115
Mixotricha paradoxa, 42–46, 53
Möbius strip, 16
monogamous tribes, 128
monotheism, 206
mood disorders, 144
morality, 128
morbus gallicus, 62
Morowitz, Harold, 233
Muhammad pellets, 72
multicellular bacteria, 22–29
multiple genomes, 44
mutualism, 96
mystical experience, 105
mystical visions during seizures, 105
myth of Narcissus, 18
National Aeronautics and Space Administration (NASA), 185
National Enquirer, 11
natural selection. *See also* evolution; sexual selection described, 94
  and genetic transfer, 89
  of groups and societies, 128–129
  and human evolution, 97
  in neo-Darwinian theory, 214
  and sexual selection, 123
  and symbiogenesis, 46
  vs. thermodynamics, 228
Navicula (diatom), 21
near-death experiences, 107–108
Nelson, Avi, 9
nerve cells, 51

nerve impulses, 53
nervous system (mind) vs. immune system (body), 24
neural Darwinism, 55, 205
neural selection, 205
neurotubules, 39
neutral relations, 24
Newtonian worldview, 21
Nietzsche, Friedrich Wilhelm, 57, 58, 60–61, 69, 198, 209
Nietzschean ecology, 220
nöosphere, 200–201
nuclear conflagration, 92–93
nuclear winter, 9
nucleated cells
  formation of, 92
  evolutionary history of, 66
nucleated organisms, 22, 34
nucleation, 136
nucleus, DNA in, 43

obligate anaerobes, 167
Oceanus Macromicrocosmicus, 158
obligate parasites, 65
operating points, 176
orangutans, 126
Origin of Species (Darwin), 157
Oscillatoria limnetica, 193
osmolyte, 188
oxic world, 169
oxidation, 192
oxidizing atmosphere, 170, 234
oxygen concentrations, 179–180
oxygenic biosphere, 192
Ozark blind salamander, 38
ozone layer destruction, 93

pain, 203
paleomammalian brain, 132
Pan bonobo, 78
Pan troglodytes, 78
paramecia, 117, 118
Paramecium aurelia, 118–119
paresis, 58–59
Parker, Geoff, 123–124
Parmenides, 224
parthenogenetic reproduction, 137
penicillin, 59, 64–65
penis anxiety, 127–128
penis, whale, 124

perception, 48
periplasmic flagella, 60
Persinger, Michael, 101, 103, 104–107
Persinger helmet, 105
personality deconstruction, 25
personal sacrifices, 7
Pert, Candace, 24–25
Phaeocystis poucheti, 188
phagocytosis, 31
Phanerozoic eon, 165
photosynthesis, 178–179, 230
photosynthetic bacteria, 179
photosynthetic blue-green bacteria, 43
Pillot, J., 64
Pillotina, 64
pineal gland, 197
pioneer species, 79
pituitary gland, 211
Planetary Biology and Microbial Ecology (PBME), 185
plants, triple genomes in, 43
plasmids, 97
Plasmodium (malarial parasite), 71
Plato, 223
pollution
  as raw material, 77, 81
  stresses caused by, 94
polygenomic organisms, 22
Pongo, 126
postmodernism, 12
pox inspection, 59
Precambrian era, 170
prejudice, racial, 36
premenstrual syndrome (PMS), 143
Prigogine, Ilya, 233
progesterone, 143
programmed death, 71
prokaryotes, 30, 167
promiscuity, 125, 128–129
propositional truth, 19
prosematic signals, 134
prostaglandins, 211
Proterozoic eon Earth, 168
protists, genomes in, 43
protoctist, 43
Protoctista Kingdom, 43
protoctist cells, 31–32
protoctist sex, 137
protoctists kingdom, 34
Pythagoras, 223

racial prejudice, 36
radiation, 93
radiation-resistant mutants, 93
radioactive elements, 82
R-complex, 130–131, 132, 133
recreational drug use, 144
redwood tree, 81
reef-building corals, 45
refampicin, 68
*Reflections on the Motive Power of
Fire* (Carnot), 227
religious experience, 105
remnant genes, 50
representation, 19
representational truth, 19
reproduction, 72
reproduction, state controlled, 149
reproductive organs, 142
reptilian brain (R complex), 130–131,
132
res extensa vs. res cogitans, 195
respiring bacteria., 92
retina, 39
Revsbech, N. P., 192
Reynolds, J. R., 144
Rhodophyta, 34
robots, 98
roll-call processes of meiosis, 115–118
Roszak, Theodore, 145

Sade, Donatien Alphonse François, 198
Sagan, Carl
death of, 9
The Demon-Haunted World, 12
DS discussion with, 102
on God and suicide, 13–14
at JPL, 154
marriage of, 2
on planets' fates, 9
public persona of, 8–9
visits Dorion Sagan in hospital, 13
Sagan, Dorion (DS)
background of, 2
magic tricks, 10
reconciliation with Carl Sagan, 14–15
relationship with Carl Sagan, 14–15
Sagan, Jeremy (brother to DS), 10
Sagan, Rachel (grandmother to DS)
death of, 11
tricks Carl Sagan, 10

Sagan, Tonio, 12
Sagdeev, Roald, 9–10
saguaro cactus, 81
Sanders, Laurie, 36
satori, 110
Schaudinn, Fritz, 62
Schrödinger, Erwin, 225, 229
Schrödinger paradox, 225, 229
Schwartz, Lawrence, 71
Schweickart, Russell L., 216–217
science
as natural selection and genetic
transfer, 89
skepticism about, 11–12
Searcy, Dennis, 37, 38
second law of thermodynamics, 224,
226–227. *See also* thermodynamics
seizures, 105
selective cell death, 206
self, 16
self, redefined, 17
self-fertilization (autogamy), 119
self-maintenance, 90
senility as spirochetal-remnant atrophy,
55
sense-of-self and autopoietic entity, 23
sensory-nervous systems, 48
set points, 176
sexual dimorphism, 126
sexual fusion, 75
sexual reproduction
cost of, 112
defined, 113
need for, 109
origins of, 43, 114
reasons for, 112
selection pressure for, 113
sexual selection
and natural selection, 123
sperm-competition, 126
Shearer, Moira, 5
silica, 81
silver nitrate drops, 64
size advantage, 133–134
Skeptical Enquirer, 11
Smil, V., 174
Smith, Robert, 125, 127
Snow, C. P., 226–227
sociosexual behavior, 135
solar energy, 230

solar gradient, creative destruction of, 234–235
solar luminosity
  and atmospheric homeostasis, 175
  and Earth's temperature, 173
Solé, Mónica, 68
solipsism, 203
Sonea, Sorin, 138, 146
space colonization, 96, 218
space-time rift, 236
speciation, 31
species creation
  coevolution with, 95
  through coevolution, 95
speculation, 48, 53
sperm competition, 124–125
sperm-competition avoidance, 126
sperm competition vs. sperm-competition avoidance, 128
spirochetal remnants, 41, 42, 46, 55
spirochete bacteria, 50, 51
spirochetes, dormancy of, 67–68 cysts
Spirosymplokos deltaeiberi, 66–69
Spirulina, 36
squirrel monkeys, 134–135
Stanhope, Philip (Lord Chesterfield), 110
Staphylococcus bacteria, 147, 151
steam engines, 84
stellar evolutionary theory, 164
Stentor, 113, 118
Stentor coerulens, 146, 149
Stephanonympha, 34
steroids, 143, 144
Stigmatella (multicellular bacteria), 24
Stolz, John F., 185
striatal complex (R complex), 130–131, 132
stromatolites, 136, 170
Stylonychia, 118
sulfur cycle., 186
sulfur dioxide, 188
sulfur dioxide readings, 189
sulfur trioxide, 188
supercosm, 89
superkingdoms, 33
superorganisms, 138, 211–213, 215
superwoman role, 6–7
surface temperature regulation, 175.
  See also atmospheric homeostasis

survival by learning, 78
Sutherland, J. L., 45
sweet memories, 54
symbiogenesis, 31, 44, 46, 73, 241
symbionts, 45
symbiotic leaps, 98
synapses, 54–55
Synapsida, 132
synaptic densities, 205
synchronicity, 212
synthetic isoprenoids, 81
syphilis
  and AIDS, 63
  in AIDS patients, 62
  of Nietzsche, 58–59, 60–61
  resistance to, 63–64
  spread of, 61–62
Syphilus sive Morbus Gallicus (Fracastoro), 62

taste buds, 39
taxonomy, 29
Teal, Tom, 68
technological revolution, 76
technology
  as ancient activity, 77
  as ancient ecological cycles of procurement, 83
  learning of, 78–79
  as part of nature, 82
Teilhard de Chardin, Pierre, 200–201
teleology, 177
temperature homeorrhesis of biosphere, 177
temperature modulation on Earth, 213–214
temperature on Earth, 231
temperature regulation, 161–162
teosinte, 95
termites, 44, 85
tertiary syphilis, 65
Tetrahymena, 201
The Antichrist (Nietzsche), 61
The Case of Wagner (Nietzsche), 61
thecodonts, 132
The Demon-Haunted World (Sagan, Carl), 12
The Descent of Man and Selection in Relation to Sex (Darwin), 125
The Lives of a Cell (Lewis), 36–37

The Proceedings of the National Academy of Sciences, 53
The Proceedings of the National Academy of Sciences., 53
The Red Shoes, 5
thermodynamics vs. natural selection, 228
Thermoplasma acidophila, 38
thermostasis of Earth, 181
The Way of All Flesh (Butler), 84
tholins, 9
Thomas, Lewis, 36–37
thought, 48, 53
three-domain system, 33
Thus Spoke Zarathustra (Nietzsche), 61
tick arthritis, 60
time direction, 225
tissue-level multicellularity and meiotic sex, 119
Todd, Neil, 99
Treponema pallidum, 59, 60, 62, 64, 65
Treponema spirochete bacteria, 46
Trichomonas vaginalis, 46
Trichonympha, 118, 119
triune brain, 131
trophic relation, 24
trust, 10
Twilight of the Idols (Nietzsche), 61
two-kingdom system, 29, 32

UFOs, 106–107
ultraviolet light, 114
Umwelt, 131
unconscious mind, 132
unit of selection in Darwinian evolution, 21
University of Massachusetts-Amherst, xv, 77, 83
Urban VIII, Pope, 197

van Leeuwenhoek, Antoni, 29, 52
venereal disease, 51

Venus
atmospheric conditions on, 178
carbon dioxide atmosphere, 231
greenhouse gasses, 9
Vernadsky, Vladimir Ivanovich, 81, 182, 199, 200–201, 231
Victoria, Queen of England, 144
Villefranche-surmer, 94
Virchow, Rudolf, 142–143
viruses, classification of, 33
visual image processing, 91
vitamin-craving yeast, 150–151
Voltaire (François-Marie Arouet), 198
von Humboldt, Alexander, 200
von Lewenheimb, Sachs, 157–158
von Linné, Carl (Carolus Linnaeus), 30
Vonnegut, Kurt, 149
von Uexküll, Jakob, 131

Walker, J. C. G, 159
Wallin, Ivan E., 44, 45
Wassermann blood test, 64
Watson, Andrew, 175, 176, 179–180
Watts, Alan, 17
Weinberger, Casper, 9
wetlands, 33
What Is Life? (Schrödinger), 229
Whale penis, 124
Whole Earth Catalog, 157
Wicken, Jeffrey, 225
Wilde, Oscar, 220
Wile, Udo J., 62
Williams, Janet, xv, 101
Wilson, E. O., 101
wings, 91

Xenophon, 182
yaws, 60
yeasts, 52

***Zany*, 148**
Zea mays (corn), 95

essay sources

"Red Shoe Conundrum" originally appeared in *A Hand Up: Women Mentoring Women in Science*, edited by Deborah C. Fort (The Association for Women in Science, 1993).

"The Truth of My Father" originally appeared as "Partial Closure: Dorion Sagan reflects on Carl" in *Whole Earth*, Summer, 1997.

"All for One" was first published as "The Beast with Five Genomes" in *Natural History*, June, 2001.

"Swimming Against the Current: Harold Kirby and the Calonymphids" originally appeared in a somewhat different form in *The Sciences*, January/February 1997.

"The Uncut Self" originally appeared in Volume 129 of the *Boston Studies in the Philosophy of Science*, Alfred I. Tauber's Symposium "Organism and the Origins of Self," Kluwer Academic Publishers, Boston.

"From Kefir to Death" originally appeared in *How Things Are: A Science Tool-Kit for the Mind*, edited by John Brockman and Katinka Matson (William Morrow and Company, New York, 1995).

"Spirochetes Awake: Syphilis and Nietzsche's Mad Genius" originally appeared as "Syphilis and Nietzsche's Madness: Spirochetes Awake!" in *Dædalus*.

"Speculation on Speculation" originally appeared in *Speculations: The Reality Club*, edited by John Brockman (Prentice Hall Press, Englewood Cliffs, NJ, 1988).

"The Atmosphere as the Circulatory System of the Biosphere—The Gaia Hypothesis" was finally accepted by Stewart Brand for the *CoEvolution Quarterly*, Summer, 1975.

"Gaia and Philosophy" originally appeared in *On Nature*, edited by Leroy Rouner (University of Notre Dame Press, 1984).

"The Global Sulfur Cycle and *Emiliania huxleyi*" originally appeared as "Sulfur: Toward a Global Metabolism" in *The Science Teacher*, January 1986.

"Descartes, Dualism, and Beyond" first appeared in *Quark* (Barcelona) April-June 1996.

"What Narcissus Saw" originally appeared in *Speculations: The Reality Club*, edited by John Brockman (Prentice Hall Press, Englewood Cliffs, NJ, 1988).

"The Riddle of Sex" originally appeared in *The Science Teacher*, March 1985.

"An Evolutionary Striptease" originally appeared as "The Evolutionary Striptease" in *Doing Science, The Reality Club 2*. John Brockman, ed. Prentice Hall Press, NY (Prentice Hall Press, Englewood Cliffs, NJ, 1991).

"Vive la Différance" appeared in a different form as "Gender Specifics: Why Women Aren't Men" in the Sunday *New York Times*, June 21, 1998.

"Candidiasis and the Origin of Clowns" appeared in the inaugural issue of *New England Watershed*, October, 2005.

"The Pleasures of Change" was published in *The Forces of Change: A New View of Nature*, (Washington, D.C: National Geographic, 2000).

"Welcome to the Machine" first appeared as "Second Nature: Welcome to the Machine" in *UMASS*, fall, 1999.

Parts of "The Transhumans are Coming" appeared in *Microcosmos: Four Billion Years of Microbial Evolution* (see p. 328), by Lynn Margulis and Dorion Sagan (Summit Books, 1986), and in "Gaia and the Evolution of Machines," by Dorion Sagan and Lynn Margulis (*Whole Earth Review*, volume 55, 1987).

"Alien Enlightenment: Michael Persinger and the Neuropsychology of God" first appeared in *If I Had Known it was Harmless I Would Have Killed it Myself*, one of a four-part avant-garde magazine called *Black Box* supported by the Danish Ministry Culture, 2003.

"Prejudice and Bacterial Consciousness" first appeared in *New England Watershed*, April-May, 2006.

"Power to the Protoctists" was originally published in Earthwatch Journal, 1992 as "Rethinking Life on Earth. The Part: Power to the Protoctists."

Sciencewriters Books

scientific knowledge through enchantment

Sciencewriters Books is an imprint of Chelsea Green Publishing. Founded and codirected by Lynn Margulis and Dorion Sagan, Sciencewriters is an educational partnership devoted to advancing science through enchantment in the form of the finest possible books, videos, and other media.

Other Sciencewriters Books available:

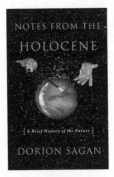
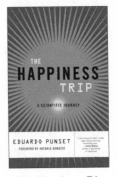